UNIVERSITY LIBRARIES
THE UNIVERSITY OF TEXAS
AT BROWNSVILLE
PRAESIDIUM
DISCIPLINA
CIVITATIS

THE
NEXT
BOOM

Books by Jack W. Plunkett

Plunkett's Advertising & Branding Industry Almanac
Plunkett's Airline, Hotel & Travel Industry Almanac
Plunkett's Almanac of Middle Market Companies
Plunkett's Apparel & Textiles Industry Almanac
Plunkett's Automobile Industry Almanac
Plunkett's Banking, Mortgages & Credit Industry Almanac
Plunkett's Biotech & Genetics Industry Almanac
Plunkett's Chemicals, Coatings & Plastics Industry Almanac
Plunkett's Companion to the Almanac of American Employers
Plunkett's Consulting Industry Almanac
Plunkett's E-Commerce & Internet Business Almanac
Plunkett's Energy Industry Almanac
Plunkett's Engineering & Research Industry Almanac
Plunkett's Entertainment & Media Industry Almanac
Plunkett's Food Industry Almanac
Plunkett's Health Care Industry Almanac
Plunkett's InfoTech Industry Almanac
Plunkett's Insurance Industry Almanac
Plunkett's Investment & Securities Industry Almanac
Plunkett's Nanotechnology & MEMS Industry Almanac
Plunkett's Outsourcing & Offshoring Industry Almanac
Plunkett's Real Estate & Construction Industry Almanac
Plunkett's Renewable, Alternative & Hydrogen Energy Industry Almanac
Plunkett's Retail Industry Almanac
Plunkett's Sports Industry Almanac
Plunkett's Telecommunications Industry Almanac
Plunkett's Transportation & Logistics Industry Almanac
Plunkett's Wireless, Wi-Fi, RFID & Cellular Industry Almanac
The Almanac of American Employers
The Next Boom

THE NEXT BOOM

What You Absolutely, Positively Have to Know
About the World Between Now and 2025

Jack W. Plunkett

BizExecs Press™

The author has used exhaustive efforts to locate and fairly present accurate data. However, when using this book or any other source for business and economic information, the reader should use caution and diligence by conducting further research where it seems appropriate. The author, editors and publishers assume no liability, beyond the actual payment received from a reader, for any direct, indirect, incidental or consequential, special or exemplary damages, including loss of income or profits, and they do not guarantee, warrant or make any representation regarding the completeness, accuracy, or merchantability of this material or the results of use of this book. As of the date of publication, the author owned no interests in the companies named or described herein other than Plunkett Research, Ltd. We wish you success in your endeavors, and we trust that your experience with this book will be both satisfactory and productive.

BizExecs Press™
An imprint of Plunkett Research, Ltd.™
P.O. Drawer 541737
Houston, Texas 77254-1737 USA

Phone: 713.932.0000
Fax: 713.932.7080
Internet: www.plunkettresearch.com

Purchases in bulk quantities are available for educational, corporate, association and conference use.

ISBN-10: 1-59392-202-7
ISBN-13: 978-1-60879-999-2
eBook ISBN: 978-1-60879-901-5

For Martha,
and in memory of friends and loved ones we recently lost:
Barbara; Jim and Vanna; Mac; Ralph and Tina

— Table of Contents —

Part Three - Things

— Tables —

Chapter One

Chapter Two

Chapter Four

Chapter Nine

— introduction —

A bumper sticker popular in West Texas during the oil bust of the early 1980s went something like this, "Please God, just give me one more boom—I promise not to blow it this time."[1] Today, millions of people around the world may be having similar thoughts. Fortunately, the foundations of the "Next Boom" are already being laid, and people who understand the massive changes that will bring that boom about will benefit greatly.

The Three Building Blocks of the Next Boom

The recipe for the Next Boom has been written: expanding markets, growing consumer demand in emerging nations, and motivated entrepreneurs combined with breakthrough technologies. The raw ingredients are already merging and taking shape like yeasty dough rising slowly in a baker's oven, even if it seemed impossible to imagine during the dismal bust we recently endured. The core focus of this book is that three powerful platforms have tremendous synergy that will boost the world of business while bringing stunning global changes during the 2011-2025 period. These vital building blocks include:

1) A soaring global population.
2) Sweeping changes in consumers, demographics and education.
3) Emerging technologies, centered on health care, wireless communications, biotechnology, nanotechnology and energy.

The Next Boom and its underlying causes will have a deep, evolutionary effect on all of us. For example, the next billion consumers are right around the corner (including the next 40 million Americans). The U.S. population was expanding at the rate of one person every 11 seconds, as of mid-2010,[2] and you will soon see headlines proclaiming that the world's population has reached the 7 billion mark.

But First, What Launched the Last Boom?

During the final months of 1982, one of the most amazing eras of economic growth in history took off with a burst of strength, stamina and hope like Lance Armstrong charging up a mountain in the *Tour de France.* The United States was entering an era of economic expansion that endured (with occasional slow downs) until 2007, but you certainly couldn't tell it from the gloomy looks on Americans' faces. In the early 1980s, consumers were generally careful with their money. Jimmy Carter had left the White House the year before, losing to Ronald Reagan's campaign concepts of supply side economics and his warnings about the "misery index," which is the sum of the rates of inflation and unemployment. People were miserable indeed. Jobs were scarce, and the American unemployment rate reached 10.8% on December 1, 1982. People were thrifty; many stayed home, watched TV or read books. When some families wanted to splurge, they went to the movies. Escapism, introspection and grand feats of courage were the order of the day in entertainment. The film *Gandhi* won the Academy Award for best picture. *Falcon Crest, Dynasty, Three's Company* and *The Love Boat* were hot on TV. James Michener's novel *Space,* about the early heroes of space exploration, and Leo Buscaglia's feel-good guide *Living, Loving and Learning* were among the most popular books, while John Updike won a Pulitzer Prize for *Rabbit is Rich.*

The economy was lackluster as 1982 began. However, as the year progressed a fire started to glow in the minds of the nation's investors, financiers and entrepreneurs that soon erupted into a wildfire that took 25 years to extinguish: the Great Boom. New technologies and brilliant business startups were about to change the world, creating massive amounts of global wealth in the process. Information technology had whetted the appetite of a new generation of entrepreneurs who launched companies in 1982 that would become the foundation of the technology boom. A representative list of these firms should include Electronic Arts, Adobe, Autodesk, e-Trade, Compaq (which, along with hard drive maker Conner Peripherals, became one of the fastest startups to reach $1 billion in annual sales while making global investors acutely aware of the potential high returns of well-placed venture capital), LucasArts, Lotus Software, Sun Microsystems and Symantec. After a promising start, personal computing was going mainstream. While Atari and Radio Shack had sold small PCs in the late 1970s and the Apple II had gone on the market in 1977, it was 1982 that saw the PC market really begin to take off, bringing thousands of companies that sold software, printers, accessories, games or computer services along for the ride. The Commodore 64 personal computer was launched at the groundbreaking price of $400, with an equally groundbreaking 64 kb of RAM memory. The IBM PC had been on the market briefly, since 1981, using an operating system called MSDOS. This software, incidentally, was co-written by Bill Gates, who went on to become one of the richest people in the world. Soon, 1983 saw the launch of Novell, AOL and Intuit on the tech side, and Sam's Club started its crusade to revolutionize warehouse-style retailing. By 1984, the Apple Macintosh hit the market and desktop publishing and computerized graphic design took off. Meanwhile, PC owners upgraded to hard drives from floppy disks for their storage needs, which made the usefulness of the personal computer virtually unlimited.

The Great Boom had a lot of help. Venture capital, leveraged buyouts, private equity, deregulation, immigration, globalization, population growth, communication technologies, huge increases in the size of the global middle class, and the massive, well-educated Baby Boom generation in America and elsewhere combined to create near-perfect conditions for global stock markets. The Dow Jones Industrial Average of stocks (the "Dow") grew from 1,100 in February 1983, after a long, stagnant period, to 5,000 in November 1995, 10,000 in March 1999 and 14,000 in mid-2007. The International Monetary Fund ("IMF") reports that world trade grew by an average 7.1% yearly from 1991-2000. Those of you who know the "rule of 72" [3] realize that a growth rate of this magnitude was enough to double world trade over a 10 year period. While a few people warned of "irrational exuberance," consumers went on a buying binge unrivaled in the history of the world. Sales of automobiles, houses, home furnishings, jewelry, apparel, electronic games, meals in restaurants, airline tickets and room nights in hotels all enjoyed amazing, long-term rates of growth. Eventually, that growth was equaled in the world of debt as interest rates fell and loans became astonishingly easy to obtain: credit card balances, car loans, home mortgages and corporate debt soared like there was no possible limit to the ability of borrowers to repay.

The dotcom era was born and technological innovation seemed endless. By 2000, unemployment in America had dropped to as little as 3.8%, and employers faced true worker shortages. Real GDP (adjusted for inflation) for the world grew by 2.8% yearly for 1991-1998 and 3.9% yearly from 1999-2006. Growth rates in emerging countries were much higher. This was the boom that lifted hundreds of millions of people out of poverty throughout the world. When you consider the magnitude of the Great Boom, which officially peaked near the end of 2007,[4] it becomes less surprising that it was followed by a significant correction: the Great Recession which devastated so many people and organizations in 2008 and 2009.

Today, Long-Term Trends Point to Another Powerful Boom

If I had to describe my point of view as an author, I would say that I am a "pragmatic optimist." I am fully aware of the multitude of obstacles lying in the path of economic growth, but I believe those obstacles will be surmounted effectively. I developed the idea of this book in early 2009. The world had been bombarded with abysmal news—stock markets spiraling downward, foreclosures and bankruptcies mounting and unemployment soaring. Nonetheless, I knew instinctively that the gloom would eventually be followed by good times. Tight credit markets, such as those of 2008-2009, always ease. New waves of opportunity and prosperity course through the world of business and industry from time to time, always followed by a correction, mild or otherwise. I began to imagine what the positive side of business news would look like in coming years. From recent work we had been conducting at Plunkett Research, Ltd., I mentally plucked the mid- to long-term trends that were going to aid a recovery and eventually lead to the Next Boom. A strong picture began to take shape in my mind, and I realized I had the basis for a book. I outlined a succinct list of trends that will soon be positive for the economies of the U.S. and the world overall. To me, these trends are nearly irresistible. While they may accelerate or slow, depending on external influences such as war, interest rates, regulatory environments, terrorism or widespread disease, they are not likely to be stopped. In addition, I believe that there has been a sea change in the attitudes of voters and consumers—today there is a rising demand in America and many other nations for more efficient, effective government, better public education and job creation, while such challenges were easy to ignore or brush aside during the heady days of the Great Boom. This change in attitude and focus has the potential to be very positive for the world of business.

While it is impossible to foresee the length, extent and exact timing of the Next Boom, vital factors are combining that have the potential to launch a very exciting period of economic growth. Demographic changes and population growth patterns point to important new consumer markets. Meanwhile, technology continues to move ahead boldly and the Next Boom will follow. The world's brightest strategists are already looking ahead to the next billion wireless subscribers, the next billion car owners, the next billion Internet users and the next billion members of the global middle class. Information technology has only begun to enhance businesses of all types, drive the delivery and usefulness of information in real time, and make industries more efficient.

There is a lot to be optimistic about if you know where to look. I'm not suggesting that you look at the world through rose-colored glasses. Instead, I believe that you will benefit most from the state of the world today by understanding the changes that are occurring around you in order to prepare yourself for the Next Boom. Economic booms and busts have been recurring in cycles since organized trade and commerce first emerged in the earliest days of civilization. October 2009 marked the 80th anniversary of Black Friday: the great stock market crash of 1929 and the key that unlocked the Pandora's Box now called the Great Depression. Here's a bit more U.S. recession history: July 1990 began a 10-month recession. April 1980 saw the start of a 30-month recession. The oil embargo crisis launched a 24-month recession in April 1973. Slowdowns, recessions and burst bubbles occur on a regular basis, including an abrupt halt to business after 9/11. The Next Boom is coming, despite the Great Recession of recent years, and it will be an exciting period of solid global growth that will create immense opportunities for investment, business formation, innovation and job creation. Yes, there will be setbacks and challenges, but the Next Boom is already starting to roll down the track—eventually, it will reach full speed.

How to Use This Book

This book is organized into four sections:

Part One: People (demographics and consumers)
Part Two: Places (global trade and rising middle classes)
Part Three: Things (innovation, always-on/always getting smaller, energy and health care)
Part Four: Change (generations and education)

As you will see, the "three building blocks of the Next Boom," including the upcoming, powerful effects of advancing technologies, are woven throughout these sections of the book. The final chapter is an epilogue that includes a scenario of what world news might look like on January 1, 2026 and a very brief discussion of "what could go wrong?"

This book is not intended to be scholarly. Likewise, it is not intended to cover all possibilities, global challenges or points of view, or to prescribe answers to all problems and needs. Instead, it is designed to be an easy-to-understand and easy-to-use detailing of the obvious and not-so-obvious, written in lay terms that will not require any particular expertise or background on the part of the reader. I am assuming that you have a reasonable interest in societal changes, technological developments and the forces that are shaping the world around you, not necessarily that you have an extensive background in business, finance or economics.

Research Tools, Discussion Guide, Multimedia and Interactive Links

The book includes tables of vital data to help you understand and analyze the numbers for yourself. In addition, the end of each chapter includes a list of "Internet Research Tips," that will make it easy for you to conduct additional research at some of the world's best online resources. For those of you reading *The Next Boom* on an Internet-enabled eBook reader, it is intended to be a highly interactive, multimedia resource. The lists of research tips include the Internet addresses of my web-based video introductions to each chapter, as well as links to numerous related videos, guides and discussions published by other organizations. If you are reading the book in printed form or on an eBook platform that lacks connectivity, you can enjoy the lists of links by entering the Internet addresses into the browser of your favorite computer. You will also find a reading group guide in the back of the book and posted to *The Next Boom's* page on my company's web site, www.plunkettresearch.com/nextboom. To participate in online discussions, you are invited to join our "The Next Boom" group on LinkedIn, or you can "like" our The Next Boom page on Facebook.

A Last Bit of Business

No one can consistently, accurately predict the future. This book was not designed to be used as specific investment advice or a crystal ball that perfectly illuminates your personal future. Instead, it was written to be a set of ideas, an analysis of global trends and synergies and a thought-provoking platform to help you do further research and make your own decisions. I have attempted to provide accurate data and analysis. Nonetheless, numbers and situations can change quickly in the worlds of business, finance, industry and government. In addition, forecasts of population, world trade and other important statistics, such as those quoted in this book, are frequently changed by the organizations that issue them. You should use caution and look for the latest data prior to making any decisions.

This book will be most useful when you keep two definitions in mind: First, when I say "near future," I mean approximately 2013 through 2025, but this is not intended to be an exact phrase. Next, when I use the phrase "Next Boom," I mean a significant, multi-year era of growth in GDP that is positive for most of the world's modern and emerging economies. Please bear in mind that the boom will most likely have periods of interruption, correction or crisis that are painful. For example, the Great Boom of 1982 to 2007 was interrupted intermittently by the Black Monday stock market crash of October 19, 1987, when the Dow stock index fell 508.32 points, losing 22% of its value; the savings and loan crisis of 1989; the dotcom crash of 2000; and the terrorist attacks of September 11, 2001. Nonetheless, America and most of its major trading partners recovered quickly after each setback. Looking back, this period was indeed a Great Boom of 25 years in length.

In closing this introduction, I will relate a story about one of the people referred to in this book's dedication. James Lorie was a close friend with whom I spent many long and entertaining evenings prior to his death in 2005 at age 83. Despite his advanced age and obvious frailness, he remained active and influential in the world of business, almost to his last days. We would meet for long dinners at his mountaintop home in Tesuque, New Mexico or dine in my house outside Santa Fe while a *pinon* wood fire roared in the fireplace. On summer evenings, we would often be joined by a group of friends to dine outdoors at one of Santa Fe's restaurants where we would listen to Jim tell his stories. Jim had been a dean and professor at the Graduate School of Business at the University of Chicago. He served on the boards of organizations like Merrill Lynch and NASDAQ, and his books and research are among the foundations of modern finance. Jim was particularly famous for analyzing the past in order to better understand the future. He could talk for hours, but Jim was also known for short sentences that were to-the-point. One evening, at a crowded dinner table, we had a lively conversation about business and what would happen in the near future. This particular conversation went on too long. In his highly effective manner, honed through decades of chairing panels, committees and meetings, Jim abruptly moved us along to another topic by stating the obvious, "There are no guarantees about the future." I say the same to you now. About the future and the Next Boom, there are no guarantees. Nonetheless, there is good reason for optimism.

Jack W. Plunkett
Houston, Texas
2010

Join in the discussion!

- See the Reading Group Guide in the back of the book.
- Go to Facebook, search for The Next Boom.
- Join The Next Boom group on LinkedIn.

— part one —

PEOPLE

— the next boom —

— chapter one —

NEWLY BORN AMERICANS, AGING BABY BOOMERS AND WHY AMERICA'S SOARING POPULATION IS A HUGE ADVANTAGE

"I always avoid prophesying beforehand because it is much better to prophesy after the event has already taken place."

-Winston Churchill

Sometime during the second half of 2010, a special person, a portent of the future, arrived in the United States of America. Statistically speaking, this arrival was probably a newborn female. She became the 310 millionth living resident of the U.S., a new milestone in the relentless upward march of America's population. As you will see, this is a strategic advantage and a source of great envy in many other nations. Overwhelmingly, the odds show that she was likely Hispanic, a ballooning sector that will account for one out of four Americans by the time she is 27 years old. Born too late to be part of Generation Y, she is a member of a new cohort, the Diversity Generation, which by 2020 will total 81 million young people. She could easily enter a career as a home health aid, based on future workforce trends. However, if the currently dismal public education system doesn't fail her (and promising reforms are gaining traction), she might become a registered nurse, an engineer, an entrepreneur or a scientist who discovers one of the vital future breakthroughs in biotechnology that will help to feed the world's soaring population. Based on the massive shifts in demographics currently underway, it isn't too hard to picture that she, or someone like her, could become the CEO of a major corporation, say Wal-Mart, or maybe a U.S. senator, or the leader of a giant, multinational financial institution.

In 2025, she will live in a world that, in many ways, is hard for us to imagine today: a world where enhanced technologies will provide immense boosts to productivity, energy efficiency, health care and food production—a world linked 24/7 by embedded communications and rapid transportation, where global trade is growing at a phenomenal rate. If America's government and educational institutions perform their jobs wisely and effectively, then she will become a young adult in an environment where the U.S. remains the world's most important economy and an enviable font of innovation and business creation. However, America will be sharing leadership to a growing extent with China while it faces intense economic competition from rapidly growing business sectors in Brazil and India, and in such advancing nations as Turkey. Africa will be a vital source of future young workers, as well as a promising market for exports, while Japan and much of Europe will suffer from aging, and in some cases shrinking, populations. These changes will be underway full steam while she is a high school student in 2025. Whoever she is, whatever she becomes, however her life, her dreams and her career evolve, she represents some of the dominant trends of the near future. She is a child of the Next Boom.

Meanwhile, we are living on a planet that supports 460 million North Americans, more than 1 billion people in each of India, China and Africa, 500 million residents of the EU and nearly 3 billion people scattered about the rest of the globe. The grand total is about 6.8 billion Earthlings. While these may sound like immense numbers, you haven't seen anything yet—populations in certain nations, including the U.S., are set to roar past these 2010 estimates like a wall of water in a tidal wave.

More people mean more demand and much larger markets for businesses. Some observers will dwell only on the massive challenges posed by such large populations. Others will make immense fortunes by solving those challenges, focusing on the opportunities, filling the needs of a booming global middle class, inventing the technologies that will move the diverse people of the world forward, and providing the infrastructure needed to sustain them. However, in order to best survive, serve or profit from this ballooning mass of mankind, you must understand the startling evolution taking place globally in demographic terms, the changing face of the typical American, and the advancing lifestyles of urban consumers in India, China and Brazil, in addition to the growth in numbers. Also, you must understand the striking differences between growing nations, such as the U.S., and shrinking nations, such as Russia and Japan. I am not implying that the U.S. or any other nation can enjoy a boom by default because of an expanding population. As discussed in later chapters, economic success on a national scale requires a carefully fostered environment in which research and innovation can flourish, entrepreneurs and businesses can create jobs, and educators can properly prepare the workforce with vital skills and knowledge, while trade and investment are supported. America has a significant advantage in the robust growth of its population, and I believe this advantage will be a powerful component of an economic rebound if America has the will to seize this opportunity and move ahead into a new era of prosperity. As you will see, the world is on the cusp of many exciting developments that together will provide a significant boost to America and many of its trading partners. A good step toward understanding these developments is a discussion of changes in population, demographics and consumer habits, as provided in Part One of this book.

Grow the Economy, Stupid

America's rapid growth of the near future will push its population to 350+ million before 2025, and 400 million around 2043.[1] This will be a vital leg under the Next Boom. In America, a rising head count will boost economic growth, fuel consumer demand and help bail out government debts and obligations. This trend will provide employers with a broad workforce while increasing both the tax base and deposits to the ailing Social Security system. In Bill Clinton's successful presidential campaign against George H.W. Bush in 1992, strategist James Carville came up with a powerful phrase, "It's the economy, stupid." This simple statement captured the attention of millions of voters and effectively painted Bush as a mediocre manager of America's future. A candidate in 2012 or 2016 might want to modify Carville's idea slightly to "Grow the economy, stupid," to help address rapidly growing concerns about federal debt and entitlement levels, which will be looming over the land like a vulture watching a wounded animal.

The United States remains an incredibly lucky land in many ways, despite current high debt levels, and despite rising competition from emerging nations like China. Part of America's good fortune lies in its population trend. The head count rises relentlessly, second-by-second, adding about 3 million people yearly since 1990. If you want to know how large the population is at any given moment, simply log onto www.census.gov, the web site of the U.S. Census Bureau, which is the agency officially charged with counting the numbers and studying the demographics of the nation. On their web site, analysts at the Census Bureau maintain two "Population Clocks" providing estimates of the total counts of the U.S. and of the world as a whole. Tick after tick, these clocks paint a very useful picture of the future.

As of June 2010, the bureau figured there was one U.S. birth every 7 seconds, which more than offset the rate of death (one every 13 seconds). Furthermore, the bureau calculated that there was one net immigrant every 36 seconds. Doing the math showed a gain of 7,854 Americans per day, enough to create a small town. In one year, it's enough to match the number of people living within the city limits of Chicago. Every 11 seconds, the clock cranks up one more person, steadily pushing America's count well past the 300 million mark that was achieved in 2006.

While you are on the bureau's web site, it is worth studying the World Population Clock as well. For the 12 months ended June 30, 2010, the bureau estimated the Earth's tally of inhabitants to have grown by 76 million. That one year's growth was an amount roughly equal to the combined populations of California, Texas and New York.

How Crowded Is America?

For a few years in the mid-2000s, I commuted by air on a regular basis from Houston, Texas to Santa Fe, New Mexico, usually via the Albuquerque airport. The Houston – Albuquerque flight is a bit less than two hours, but it nevertheless covers a lot of territory, through a vast expanse of central and west Texas, and on to central New Mexico. Two facts were apparent as I gazed out the airplane windows on those repetitive flights. First, there is a massive amount of room left in America for future multitudes to live and work. Yes, the greater Houston area may cover 600 square miles, and it looks pretty well packed from the air, but it quickly passes from sight as rolling, rural Texas countryside emerges. The next major city in the usual flight path is Austin, with about 1.7 million people in the metro area as of 2010.[2] Then there is next to nothing: hundreds and hundreds of miles of near emptiness until the aircraft abruptly descends over the Sangre de Cristo mountains to provide a first glimpse of Albuquerque which, like Austin, is a city of relatively modest size. These open expanses and mid-size cities are important building blocks of the future. America's vast size will give its growing population a wide variety of choices in terms of urban, suburban and rural lifestyles. As you will see, population density in the United States is remarkably low.

A second fact that I mulled over on those aircraft: while technologies leap forward constantly, this progress may be largely out of sight to most of the population. From the air, traveling west from Houston or Dallas has always offered a look at oil and gas wells far below on the Texas plains. Their number increased quickly in the mid-2000s as the price of oil soared. But, at the same time, wind mills sprouted beneath the flight path—thousands of them, turning Texas into America's largest wind-power generating state almost overnight. Change is constant; progress is nearly as constant, but we may not be very aware of it. Technological progress is a second, absolutely vital building block of the Next Boom, and it will enable prosperity to advance along with population growth.

1 Billion Americans?

To a stunned crowd at the American Planning Association's 2008 conference in Las Vegas, demographics expert Arthur C. Nelson made the case that America can sustain immensely larger populations, and may well need to do so in the relatively near future. He explained that by 2100 to 2120, the U.S. head count could grow to as much as 1.0 to 1.2 billion (at the high end of Census Bureau estimates), due to high fertility rates, continuing streams of immigrants and the much longer life spans of the future.[3] While population increases of this order would create intense demands on infrastructure such as energy, transportation and water supply, denser housing could easily make room for a vastly higher number of Americans. This is an important thought, when you consider the great success that some nations have enjoyed in developing prosperous but extremely dense cities. The United Nations estimated the 2010 density of the U.S. at only 33 people per square kilometer, far below the average density of all the world's land mass, excluding Antarctica, (55) and vastly below that of Singapore (7,082), South Korea (487) and Japan (336). A mammoth surge in U.S. population to 1.2 billion would create a density of about 130 people per square kilometer, about 20% higher than today's France and significantly lower than today's Germany.[4]

Growing the Tax Base

There is intense concern in the finance and investment community that the U.S. government will be forced to utilize inflation to help ease the effect of today's massive federal debts. This may or may not turn out to be true, but there is a much better method than inflation to help reduce these obligations: utilize growing tax revenues from a rising population base and a booming economy. If you have any doubt of the value of a growing, working-age population in the U.S., consider the following facts: As of the fourth quarter of 2008, the Peterson Foundation estimated that unfunded U.S. federal liabilities totaled $56.4 trillion, an amount equal to $483,000 for each American household. This figure includes Medicare benefits not covered by taxes and other contributions ($36.3 trillion), Social Security benefits similarly not covered ($6.6 trillion) and other federal government liabilities ($13.5 trillion).[5] If you are an American taxpayer, exactly how do you plan to pay your share? Federal deficits planned for subsequent years will make this look even worse. It will be a great deal easier for future administrations to find a way to fund this financial abyss if there is a much larger workforce, a thriving entrepreneurial business community creating new jobs, and a much bigger base of taxpayers.

A Brief Lesson from U.S. Population History, the Baby Boom and the Re-opening of the Door to Immigrants in the 1960s

Predicting population growth is an inexact art, and forecasts often prove to be embarrassingly far off the mark. The late Julian L. Simon wrote widely on the likely ability of the Earth to provide for an expanding number of humans. In the second edition of his book *The Ultimate Resource*, he pointed out that "The history of demographic predictions calls for humility…and teaches caution rather than fear-motivated, over-reactive policies."[6] Simon tells the story of a "Presidential Research Committee" composed of noted scientists who, in the depths of the Depression in 1933, reported to President Herbert Hoover that America "shall probably attain a [maximum] population between 145 and 150 million in the present century." The 20th Century's end on December 31, 1999 proved them basically blind to future developments, as the actual population turned out to be 279 million. From the day of their forecast in 1933, through the end of the century in 1999, America emerged from the Great Depression, battled in yet another world war and enjoyed several periods of economic boom. Meanwhile, technologies in food production and health care advanced with a resulting boost to nutrition and an extension of average life span, the Baby Boom ensued and immigration brought waves of new people in the tens of millions to enjoy the fruits and freedoms of the United States.

Future growth in America's population will not be steady. In fact, it varies widely from year to year, and many adjustments to estimates will be required over time. The state of the economy is a major factor in population growth and new household formation. For example, the lowest population gain of recent history was during the height of the Depression in 1933. Still, even during a miserable year known for hunger, poverty and unemployment, human reproduction remained a powerful factor, and the U.S. population increased 0.59% for a net increase of 741,000.

The post-war period saw the head count grow dramatically, starting in 1947, as soldiers were resettled back home and many women left the work force or their military and wartime volunteer posts. (This population boom occurred throughout much of the world.) In 1950, America's annual population gain exceeded 3 million for the first time. The "Baby Boom" was well under way, roaring ahead non-stop through 1964, resulting in today's 76 million surviving Baby Boomers. As you will see, the Baby Boom that started more than 60 years ago is about to have a significant effect on the near future.

The face of America began to change dramatically after a 1965 reform of immigration laws in America opened the doors to large numbers of new residents, immediately after the Baby Boom ended. As a result, an immigration boom ensued that, in terms of numbers, equaled the booms of the late 19th and early 20th Centuries, when newly arrived Europeans and Asians flooded U.S. cities. An important factor in the success of this 1960s renewal of immigration is that America, with its centuries-long history of growth via hopeful new arrivals from Ireland, Poland, Germany, China, Mexico, Africa and elsewhere, did a better job of assimilating its post World War II immigrants than did many other nations. Today, this trend that began decades ago is having a dramatic effect on the America of the 2000s, as second- and third-generation members of many immigrant families obtain high levels of education and push forward America's innovation, technology base and economic output. If you have trouble grasping this, check the number of Asian surnames on the roster of physicians and scientists at any major American teaching hospital.

Later, social upheaval of the Vietnam era and the soft economy of the 1970s led once again to tepid population growth until an immense bounce back starting in the early 1990s. At that point, the economic outlook was brighter and the technology boom gained traction with the widespread introduction of cheap personal computers and cellphones, along with rapidly growing adoption of the Internet. Jobs were plentiful and household formation was high, along with strong growth in births and immigration.

Immigration—Problems, Policy and Potential

Immigration has always been a fundamental component of America's growth. Many of today's immigrants are highly educated professionals. However, a large number of immigrants have little education and tend to fill low-paying, entry-level or temporary jobs, such as those in meat processing plants, landscaping and agriculture. They also fill jobs in elder care, a need that is going to soar in the near future as Baby Boomers age. U.S. labor market conditions of 2009-2010 reduced immigration because job availability was constrained.[7] When unemployment falls significantly, immigration is likely to resume in earnest.

The logistics, legislation and implications of immigration remained a politically and emotionally charged quagmire as of 2010. It is a national challenge that bears full, impartial and continuous attention. There is the ongoing need to fill low-end jobs, such as home health aides and food industry workers.

Do low-skilled Americans have the motivation and mobility required to fill these jobs, or is a supply of immigrants essential to such tasks? There is the potential to attract and encourage the immigration of highly educated scientists, engineers, researchers, doctors and entrepreneurs who could boost America's competitive advantage in science and technology. There is the possibility of boosting the economy by retaining and assimilating the most gifted of the foreign students who come to the U.S. by the tens of thousands each year to obtain advanced degrees and vital work experience, many of whom currently take their educations, useful experiences and talents back to the benefit of their home countries, particularly India and China. There is the need to resolve the problems of selected border states and communities that bear the costs and disruptions fostered by large numbers of illegal immigrants and by the lack of an effective national immigration policy.

Last, but not least, there is the need to deal with the status of approximately 11.5 million illegal aliens residing in the U.S. What is America's policy going to be? Round them up and send them home? Tolerate them? Integrate them effectively into society by creating paths to legal status for illegal aliens who qualify as desirable long term residents? Such a strategy would come with strings attached. For example, a first step could be temporary work permits for those who register with the government, file income tax returns and have no criminal background outside of their illegal entry. Current problems, particularly in U.S.-Mexico border areas, include human smuggling, drug smuggling, document fraud, trespassing, violence and a shadow workforce that avoids taxation. The illegal status of millions of immigrants strikes a raw nerve among vast numbers of American citizens, and the fact that the government is unable to control the borders is infuriating to many. There is absolutely no reason why America should settle for lackluster immigration enforcement and poor immigration policies that lead to millions of illegal residents, on one hand, and the rejection of highly skilled immigrant scientists and engineers on the other. Meanwhile, as the number of senior citizens soars in the United States over the near future, and the economy resumes serious growth, the nation will likely have a vital need for large numbers of new, legally documented, immigrant human capital during the Next Boom, including workers on the highly educated level and the unskilled level. Needed workers might come in the form of immigrants who could enter the country legally and be fully accounted for, taxed and employed under better-managed visa and guest worker programs. A primary cause of America's uncontrollable borders and large number of illegal residents is the fact that, as *The Wall Street Journal's* editors succinctly put it, "the demand for entry visas outstrips the supply…the reality is we can't secure the borders without reform that allows more legal ways to work in America."[8]

U.S. Population, Selected Years 1930-2010

(In millions; Resident population as of July 1st; Excludes Alaska & Hawaii before 1950)

Year	Total Population	Net change from previous period[1]
1930	123.077	16.616
1940	130.880	7.803
1950	151.868	20.988
1960	179.979	28.111
1970	203.984	24.005
1980	227.225	23.241
1990	249.623	22.398
2000	282.224	32.601
2010	310.233	28.009

[1] Net change from prior year shown; 1930 from 1920, etc. Change calculations for 1930 and 2010 are Plunkett Research estimates.

Source: U.S. Census Bureau

Trends That Will Fuel the Coming American Population Boom:

Population and Households:

- Total population will rise rapidly, with about 3.2 million average yearly U.S. growth over the mid-term.
- The Latino segment will show the biggest growth.
- A strong fertility rate will boost yearly births from about 4.2 million in 2010 to 5.6 million by 2050.

Immigration:

- Immigration slowed due to the Great Recession, but is likely to accelerate when economic growth resumes. During the last boom, immigration was running at about 1.5 million people yearly. It may reach as many as 2.0 million immigrants yearly by 2050.

Seniors:

- Growing life expectancy will create a vast senior population, where the 65+ years segment will swell from about 40 million in 2010 to 80 million by 2040. This means that more than one in five Americans will be what we currently refer to as "senior citizens." However, longer careers and life spans could easily move our concept of "senior" status up to 70 or 75 years of age.

The Hispanization of America

One particularly dramatic demographic change could be described as "the Hispanization of America." The Hispanic segment is growing at such a rapid rate, through both fertility and immigration, that a sea change is occurring. A major influx of a specific segment of people is not unprecedented in the history of America's development. In earlier decades, extremely rapid changes occurred due to massive arrivals of immigrants from Germany at one time (an estimated 6 million immigrants from 1840 to the beginning of World War I) and Ireland at another (by one count, the Irish were ancestors of 12% of the American population of 2006). Large numbers of Chinese immigrants changed the face of the West Coast in the mid to late 1800s, and hundreds of thousands of Vietnamese entered the U.S. and assimilated quickly, starting with the end of the Vietnam War. Today, however, and for the foreseeable future, Hispanics hold the keys to demographic change and population growth. This segment originated largely in Mexico, but has also been fueled by immigrants from Central American nations such as El Salvador and Guatemala as well as Cuba, Brazil and elsewhere. By as early as 2037, Hispanics in America will total 100 million and make up about 25% of the population.

Meanwhile, America faces an immense challenge: devising effective ways to fully integrate this Hispanic sector, a cohort of rapidly expanding size and economic importance, into society in a way that will foster future generations of well-educated, highly productive citizens. Unfortunately, this is a multi-faceted problem; many immigrants have not acquired high proficiency in the English language, and Hispanic high school students suffer from an 18.3% drop out rate, according to the National Center for Education Statistics, compared to 9.9% for blacks and 4.8% for whites. (The good news is that these drop out rates declined by about one-half from 1980 through 2008.)

Businesses and investors wishing to capitalize on trends will see this Hispanization as a ripe opportunity. To begin with, Hispanics will become more mainstream as their adoption of American lifestyles

How Hispanics are Counted

Hispanics or Latinos, as defined by the U.S. Bureau of the Census, are those people who classified themselves in one of the specific Spanish, Hispanic or Latino categories listed on the census questionnaire ("Mexican, Mexican Am., Chicano," "Puerto Rican," or "Cuban") as well as those who indicate that they are "other Spanish/Hispanic/Latino." Persons who indicated that they are "other Spanish/Hispanic/Latino" include those whose origins are from Spain, the Spanish-speaking countries of Central or South America, the Dominican Republic or people identifying themselves generally as Spanish, Spanish-American, Hispanic, Hispano, Latino and so on. Origin can be viewed as the heritage, nationality group, lineage, or country of birth of the person or the person's parents or ancestors before their arrival in the United States. People who identify their origin as Spanish, Hispanic, or Latino may be of any race. Thus, the percent Hispanic should not be added to percentages for racial categories. Tallies that show race categories for Hispanics and non-Hispanics separately are available.

Source: U.S. Census Bureau

continues along with their continued intermarriage with non-Hispanics. At the same time, their own tastes, styles and culture will have an influence on everyday American life. If you have difficulty envisioning this, consider the immense number of extremely popular "Mexican" restaurants located all across America, or wander the food aisles of any major supermarket and note the wide variety of Mexican foods and other Hispanic foods, such as those packaged under the popular Goya brand. (An Anglo living in San Antonio, Texas became quite wealthy by building his Pace brand of Mexican sauces to national prominence and later selling it to the Campbell Soup Company.)

The retail and telecommunications industries have been quick to tailor offerings to the Spanish-speaking community. Both Wal-Mart and Best Buy are installing bilingual English/Spanish signage in their stores in neighborhoods that are high in Hispanic counts, and they operate Spanish-language pages on their web sites. Today, savvy businesses know that marketing to Hispanics may involve a number of special initiatives and new services, including signage in Spanish and Spanish-speaking employees on hand. Advertising campaigns aimed at Hispanic consumers are now common, and several advertising agencies specialize in this market. Another example of an industry adapting to the Hispanic market has been financial services. Bank of America and Citibank have both designed special branches in locations where the percentage of Hispanics is particularly high. Most of these lie within Arizona, California, Florida, New Mexico, New York and Texas.

When Does the White Majority Become the Minority?

The growth of the Hispanic segment is dramatic, but it is not by any means the only big change underway. White "majorities" are rapidly becoming minorities in many cities, and that trend will continue aggressively. I recall a day in December 2005 when I made a trip to Houston's massive Galleria shopping mall, home to hundreds of stores ranging from a mainstream book store to popularly priced clothing boutiques to high end luxury retailers including Neimans, Saks and Tiffany. Weaving my way through the crowd, I was suddenly struck by the fact that, among the hundreds of shoppers I passed, I saw mainly Hispanics, Asians, Middle Easterners and African Americans. There were shoppers in modest Muslim garb, brightly colored Indian saris, African robes and turbans among mainstream people wearing hip or high fashion American clothes. But there were few white faces that made me think of English, Irish, French or German ancestry. It struck me that I was transported to the future of American shopping, where successful retailers would adapt quickly to local demographics and ethnic tastes. America's "minority" population was about 35% of the total as of 2009. In addition to America's rapid growth in the Hispanic segment, (15.8% in 2009), there are significant numbers of Asians (4.5%), and there have long been large numbers of African Americans, but their share of the total population is declining rapidly (12.3%). "White persons not categorized as Hispanic" were 65% (down from about 79% in 2000). Minority segment growth will continue to create a significant evolution in political and demographic terms. In August 2008, the Census Bureau projected that white children will become the minority in the 18-and-under age segment in 2023, and that white people of all ages will become the minority in 2042. A tipping point was expected near the beginning of 2011, when more than 50% of births would be to minorities.

Minorities are already the majority in Texas, Hawaii, New Mexico, California and the District of Columbia. The effects of the Great Re-

Projections of the Hispanic Population in the U.S.: 2010 to 2050

(In millions; Resident population as of July 1)

Year	Population
2010	49.726
2015	57.711
2020	66.365
2025	75.772
2030	85.931
2035	96.774
2040	108.223
2045	120.231
2050	132.732

Source: U.S. Bureau of the Census

cession, including temporarily slower immigration and slower new household formation, may delay the progress of this transition, but the point is clear: America is well on the road to becoming more of an ethnic melting pot than ever.

While immigration is a powerful factor in population gain, the rate, recently estimated at about 1.3 million net new immigrants in the U.S. yearly, is not necessarily constant. You might assume that immigration rates change due to federal regulations and the rate of enforcement of those laws. That is certainly correct, but immigration is also subject to the influence of the economies of the United States, and of the home country of the person who is considering emigrating. The deep, recent recession created an immediate drop in immigration, and many immigrants already on U.S. soil, legally or not, found themselves out of work. More than a few returned home discouraged.

In 2010, as it does every 10 years, the Census Bureau employed legions of people and millions of pieces of mail in attempting an accurate count of the American population. Hundreds of thousands of temporary workers were employed shuffling papers, making phone calls and knocking on doors in a last gasp effort to get people who had not responded to Census questionnaires to cooperate. The effort cost about $14.5 billion and will have a direct effect on politicians and voters alike, as the number of congressional seats allotted to each state is determined by census count.[9] You might be wondering whether illegal immigrants are counted as part of the official population. Good question. The goal of the Census Bureau is to count, and gain basic demographic knowledge of, all citizens and all residents of the nation, regardless of their legal status. Census forms are mailed to all known residential addresses, and personal calls are made on addresses that do not respond to the mail. It is easy to imagine that a household of illegal aliens won't fill out the form, and won't want to talk to a stranger carrying a clipboard and asking questions about the age, race and occupation of a building's residents. Consequently, some observers feel that illegal aliens are largely undercounted. Projections of growth in population made before the 2010 Census will eventually be recalculated and may be ratcheted down a bit.

Population—A Boom for Some Countries, a Painful Bust for Others

- In 2050, the global population may peak near 9.1 billion (about 55 million average yearly growth from 2010).[10]
- Growth will not be universal—according to the United Nations, only nine nations will account for about 50% of the increase (U.S., India, China, Pakistan, Bangladesh, Nigeria, the Congo, Uganda & Ethiopia).
- Fertility rates, the quality of health care (and the typical life span) and immigration rates vary widely from region to region.
- The effect of population changes will vary widely, creating highly negative or positive impacts, depending on locale.
- In fact, many developed economies will suffer from significant population declines.

Babies, Boomers and the Next Billion Humans

Earlier, we looked at the ways in which a growing, changing population in the U.S. will bolster the workforce, increase output, fuel demand and help create the Next Boom. When you consider the rapid rates of new births and immigration, it is easy to envision the economic activity of the America of the near to mid-future: a growing workforce and consumer base, new households being formed, continued high demands on public and private education, as well as new residential neighborhoods being built alongside new schools, shopping centers and workplaces.

Now, in contrast, imagine a nation with a very low birth rate and little immigration, combined with a long life span and excellent health care. These factors paint a picture of an aging nation brimming with senior citizens, but suffering from a lack of children, little need for new construction other than buildings created under government stimulus plans, little innovation, a shrinking base of people of working age and a disproportionately high economic burden of health care and pensions for the aged. Such a nation will have a severe challenge in maintaining a workforce of sufficient size to grow the economy, fuel the tax base and care for its senior population, and may be forced to dramatically increase immigration, temporary work permits for foreigners or offshoring of as many tasks and services as possible. This is the direction that nations such as Japan and Italy are headed under current trends.

Why You Need to Understand Fertility Rates

Geopolitical analyst George Friedman hammered an important concept home in his recent book, *The Next 100 Years*, "The single most important demographic change in the world right now is the dramatic decline everywhere in birthrates. Let me repeat that: the most meaningful statistic in the world is an overall decline in birth rates."[11] There are numerous ways to analyze the number of babies born into a population each year, including fertility rates, birth rates and infant mortality. The most instructive number for our purposes is called the "total fertility rate (TFR)." TFR continues to be high in much of the lesser-developed world. In regions like Africa, nations with poor health care and low incomes produce large numbers of babies, many of whom do not live very long. What many people are unaware of, however, is the fact that the fertility rate is relatively high in the U.S. when compared to the rest of the world's most developed nations. Primarily, the idea you need to understand is that TFR is the estimated average number of births per woman. A TFR, and its effect on population growth or decline, can be established for any nation at any point in history, provided appropriate data is available.

In the most simplistic possible terms, a TFR of 2.0 would provide for population replacement if the population is one-half men and one-half women. Under those circumstances, if the average woman has two babies, and those babies live to be adults, then that average

U.S. Population Projections,
Selected Years 2010-2050

(In millions; Resident population as of July 1)

Year	Population	Net change from previous period
2010	310.233	27.799
2020	341.387	31.154
2030	373.504	32.117
2040	405.655	32.151
2050	439.010	33.355

Source: Population Division, U.S. Bureau of the Census, release date August 14, 2008

woman has produced children sufficient to replace herself plus one man when she and that man eventually die. In reality, a TFR of 2.1 to 2.2 is more likely the minimum required to avoid a declining population in a highly developed economy, where good nutrition and post-natal care lead to high survival rates among infants. TFR may change quickly, due to factors such as the strength of the economy, changes in the ethnic makeup of a population, government incentives and changes in the social attitudes of women of child-bearing age. In 2008, the U.S. fertility rate was found by the Census Bureau to be 2.1. In the mid-1950s post-war Baby Boom, the rate soared to about 3.6 from a Depression-era low of 2.2. After dropping to about 1.8 in the mid-1970s (relatively soon after the 1960 commercial introduction of the birth control pill), the rate has been climbing to the point that it was recently at or above the replacement rate. Why is this significant? Because America's TFR, when combined with its high level of immigration, will create massive growth in the population over a reasonably short period of time: the next 50 million Americans truly are right around the corner, and a wealth of positive economic activity will arrive with them.

In its "*World Population Prospects: The 2008 Revision*," the United Nations provides both a history of TFR by country and a projection through 2050.[12] The highest TFRs were found to be in very poor nations such as Yemen, a desolate and desperate place in the Middle East (not a major oil-producing state) with a TFR of 4.65. Africa's average number was 4.27, and TFRs were quite high in the least developed countries overall at 4.08. In poorly developed economies, there not only may be little birth control, but parents are also driven by economic need to produce large numbers of children in the hope that enough of them will live to help run the farm, tend the store or get jobs in major cities to help support the family. Without significant wealth or organized pensions, parents need children to support them in their old age. The birth rate tends to decline dramatically as a nation's economy becomes more sophisticated.

This Relatively Strong Fertility Rate in the U.S. Shows a Sharp Contrast to Rates in Other Highly Developed Nations

- In the UN study referred to above, estimated fertility rates were extremely low in much of Europe, at an average TFR of 1.53. Individual examples include Spain (1.56), Italy (1.41), Germany (1.34), Greece (1.41) Austria (1.41), Bulgaria (1.5), and Poland (1.29). (Exceptions may include Britain, Scandinavia and France, where TFRs appear to be trending upward.)
- Many Asian nations also have low to extremely low fertility rates, for example: Japan (1.27), South Korea (1.26), Thailand (1.85) and Singapore (1.29). (Singapore's iconic politician and long-time leader Lee Kuan Yew points out that "Unlike Japan, Singapore welcomes younger skilled and educated immigrants from many countries…These immigrants keep Singapore's economy vibrant and reverse the graying process."[13] However, the presence of very high numbers of foreign-born workers created controversy when employment declined during the recent recession.)

What about the "BRIC" (Brazil, Russia, India and China) countries that are famous for rapid economic growth? The study cited above found the world's largest nations, China and India, to have TFRs of 1.79 and 2.52 respectively. China's history of encouraging only one child per family will eventually lead to an unreasonably large, burdensome senior segment—a preference for male children also means a shortage of females. Brazil fared poorly at 1.7. Russia has a significant problem. Not only is its TFR extremely low (1.46), it also has a very short life span and high rates of alcoholism, which hurt economic output, the overall health of the population and economic growth.[14]

It is true that TFRs can and do change over time, and nations with low scores may start to trend upward. Nonetheless, for the near future, America clearly has a significant economic advantage in its growing population:

- America's upcoming generations will be reasonably well educated, with the potential to help fuel growth in GDP (gross domestic product) and high productivity. With a little luck they will also be much healthier than average Americans of 2010.
- The growing population will help America avoid the workforce shortages that will plague other developed economies in coming years.
- The next generations will help balance the rapidly growing senior segment of the population. A significant portion of the U.S. population will remain of working age, provide needed goods and services to the non-working population, and contribute steadily to programs such as Social Security and Medicare. Maintaining a sufficient, young workforce is a significant issue. For example, the U.S. Bureau of Labor estimates that aging and infirm Americans will require a 51% jump in the number of personal and home care aides over a 10 year period, from 767,000 in 2006 to 1,156,000 in 2016. The number will grow dramatically beyond 2016 as the sheer mass of seniors increases.
- The next generations will bring innovation and fresh thinking, and they will readily adapt to new technologies. More on this later.

The Next Billion Humans

Aging, shrinking, migrating populations will create unique global challenges and business opportunities:

- The next billion Earthlings will help boost the population to 8 billion as early as 2023 to 2025.
- This growth will be concentrated in a handful of nations, as discussed above. Large numbers of new people in Africa and India will create significant social, economic and infrastructure challenges.

- Demand for food, energy, water and industrial commodities will be extremely high.
- There will be nearly unlimited business opportunities for local entrepreneurs and multinational corporations as incomes grow in nations that successfully adapt to the coming boom. Such opportunities will include products that serve the needs of local or ethnic populations; the expansion of personal services, health care and personal care products; housing; food products; restaurants, entertainment, sports and other discretionary items; financial services; telecommunications and transportation. Companies that do best in the coming global growth environment will be those that are innovative; quick to react to trends and shifts; integrate their marketing, communications, and customer care in a way that takes full advantage of Internet and wireless platforms; and focus on providing lasting, quality products, responsive service (online, via the telephone and in person) and prices that clearly represent very high value for the money spent. This may sound like a lot to ask, but that's what it's going to take in the extremely competitive environment of the near future.

Senior Citizens and the Dependency Ratio

The best way to fully grasp the implications of aging populations is to study a simple statistic known as the "dependency ratio" (DR), that is, the portion of a population that is likely not to be working compared to the number of people of working age. (DR may be calculated in several ways. For example, it is common to include both children below working age and senior citizens in the likely pool of people "dependent" on the working segment of the population.) For the purposes of this chapter, since we are discussing the challenges caused by the rapidly growing number of senior citizens, we are going to look at DR based solely on the number of people aged 65 or more, divided by the number of persons who are in the labor force. Such a DR has been carefully estimated for a group of nations with well-developed economies, the countries that are members of the OECD (Organisation for Economic Co-operation and Development), as shown in its *OECD Factbook 2009*. In 2000, the dependency ratio was 27.2% on average among the OECD nations. That means that there were a bit

Projections of the Population by Selected Age Groups for the United States: 2010 to 2050

(Resident population as of July 1. Numbers in millions)

	Year		
Age Group	2010	2030	2050
All (Total Population)	**310.233**	**373.504**	**439.010**
Under 18 years	75.217	87.815	101.574
18 to 64 years	194.787	213.597	248.890
65 years and over	40.229	72.092	88.547
85 years and over	5.751	8.745	19.041
100 years and over	0.079	0.208	0.601

Source: U.S. Bureau of the Census

fewer than three persons of 65+ years of age for every 10 people in the labor force. By 2010 however, many countries began showing signs of future strain as the average DR within the OECD rose to an estimated 31.3%. That was only the beginning of long-term troubles. By 2050, estimates paint a dark picture, with an average DR of 62.3% in these nations. In other words, there will be 6.2 people of 65+ years of age for every 10 people likely to be in the workforce. Among nations forecast to suffer the most devastating DRs in 2050 are Germany at 73.9%, Spain at 91.3%, Poland at 83.1%, Korea at 91.4% and Japan at 94.9%. Italy tops the chart with a forecast DR of 98.5%. That means that there will be only slightly more than one member of the labor force for each senior in Italy. (These figures ignore the fact that people are going to be increasingly likely to work later in life than they did in the past. I believe that the dependent age will soon be moved up to 70. More on this thought later.)

America's DR is forecast to grow from 26.0% in 2010 to 50.3% in 2050. While this number is undesirable, it nonetheless shows that America will be in much better shape than other highly developed nations, which will give it a considerable economic advantage.

Selected Population Shifts, 2050

- U.S. Population: 439 mil., +43%
- EU Population: 512 mil., +3%
- Japan Population: 110 mil., -12%

Thoughts about Business Opportunities Created by Aging Populations

Consider the following trends among the 76 million currently surviving Baby Boomers that will occur as they age:

- They will need specialized services and infrastructure. "Universal Design" will be stressed over the long term. That is, there will be more demand for everyday items designed to provide better access to people with limited physical strength and agility—ranging from crosswalks with longer timing on green "walk" lights, to wider doors that accommodate wheelchairs, to kitchens and bathrooms that have easier-to-access appliances and cabinets, as well as fixtures that are easier to handle for arthritic hands. A good example is today's trend of placing clothes washers and dryers on pedestals so they can be reached with less bending over. Remodeling and personal services that enable "aging in place" will be in high demand.
- Baby Boomers will be anxious to travel. Specialized travel services that make it easier, more exciting, affordable or more meaningful to travel will be in big demand. For example, group tours that take active seniors to do volunteer work in low-income nations may be in high demand.
- Many of these seniors will have reasonably high incomes and household wealth, requiring intelligent, honest and well-designed financial planning and delivery of senior-specific financial products such as reverse mortgages. Personal financial assistance, such as helping seniors to monitor and pay bills, may become a growth industry. (Yes, the wealth of many seniors dropped precipitously during the Great Recession, thanks to reduced home and stock portfolio values, but asset values will grow, and people who are nearing retirement age will be saving and investing for the future at an increased rate.)
- Many mentally sharp Baby Boomers will seek new careers and continuing education.
- The market for hobbies will boom.
- Inevitably, this segment will have immense health care and personal care needs as part of a longer average life span.

Seniors 65 Years+ as a Percent of the Population, and Dependency Ratios, in BRIC Nations

Year	Seniors as a % of Overall Population	Dependency Ratio
	Brazil	
2010	6.7%	13.6
2030	12.1%	24.4
2050	18.8%	39.2
	Russia	
2010	12.6%	23.8
2030	18.9%	37.4
2050	23.8%	52.9
	India	
2010	5.3%	13.9
2030	8.8%	20.6
2050	14.5%	32.6
	China	
2010	8.4%	13.6
2030	16.2%	30.2
2050	23.7%	48.0

Source: Organisation for Economic Co-operation and Development (OECD)

Aging in the BRIC Countries and Hordes of Young Workers in Africa

A young workforce and low DR will be of considerable economic advantage to two of the four BRIC nations: India and Brazil. Russia, however, will face a considerable challenge due to the size of its older population segment. China will also face challenges. Its senior segment is growing quickly, and the nation now faces a significant burden in developing social services and support that can provide adequate health care and pensions. Meanwhile, the African continent may hold the world's major supply of future workers, thanks to a young median age and the influence of a high fertility rate. The McKinsey Global Institute estimates that Africa will be the home of 1.1 billion people of working age in 2040.[15]

Feast or Famine? Shortages or Progress?

At current growth rates, the world's masses will increase by 1 billion people in about 13 years. To put this number in perspective, that is roughly the same as adding to the global population in an amount equal to the current population of India. It is about the same as adding the number of people currently living in America, Canada, Mexico and the European Union combined.

There have long been commentators who want us to be terrified of a world teeming with a soaring number of inhabitants who face universal shortages and potential starvation. The most noteworthy may have been Thomas Malthus, a British political economist who lived from 1766 to 1834 and famously forecasted that the world's population growth would inevitably outpace agricultural production, potentially forcing subsistence lifestyles or famine. This is a recurring theme among many writers. A more modern doomsayer is James Howard Kunstler, author of several books, including *The Long Emergency*. Kunstler makes a case for drastic declines in living standards and a collapse of suburbia due to shortages of such resources as petroleum. Paul Ehrlich's dire predictions in his book *The Population Bomb* led to a large number of people endorsing a policy of "zero population growth." More famously, an organization known as the Club of Rome issued a report in the early 1970s called *The Limits to Growth,* with extremely gloomy predictions of a global population expanding to the point that it exceeds the world's ability to support its inhabitants.

Nonetheless, the modern world has a remarkable record of capitalizing on advances in technology, agriculture and transportation, combined with intermittent improvements in political stability, to support a growing population. Food shortages are largely a thing of the past in nations with modern distribution infrastructure, effective governments and reasonably modern economies. Agricultural production boomed in recent decades on a worldwide basis, fostered by a "green revolution" based on technological advances in farming and seed genetics, while food availability for consumers soared along with better food processing and logistics. India has evolved from a nation of hunger to a frequent exporter of food. The Ukraine, Africa, Argentina and Brazil have immense potential capacity to expand agriculture in order to feed a booming world population.[16] China has great agricultural potential, but faces significant water supply challenges. As you will see in another chapter, agricultural biotechnology has the potential to spark a renewed, 21st Century green revolution, bringing farm output to a much higher level over the near future, while better farming technologies, systems and practices can help overcome the myriad challenges posed by the need to fertilize and irrigate fields as well as the need for efficient farm-to-market logistics.

Many writers today are concerned about possible scarcities of fresh water, petroleum, electricity and food, as well as the increasing strain that growing populations and their rising affluence put on animal habitats, air quality and water quality. These are valid concerns, and it is neither my intent to dismiss them casually nor to attempt to discuss them all in depth in this one volume. If history is any guide, many of these concerns will eventually be addressed to a large degree by improvements in technology, more efficient services and infrastructure, improved communications, increased scientific knowledge, better supply chains and evolving conservation. Entrepreneurs, investors and organizations that take the lead in developing, commercializing and distributing these evolving technologies and services will have immense opportunity to profit. In the end, the biggest infrastructure challenge of all may lie in providing and sustaining sufficient water for the personal, industrial and agricultural needs of 9 billion people, and many observers are now stating that "water is the next oil." Even for this precious resource, there is room for considerable optimism. In *Water*, Steven Solomon's epic history of the world's water supply and demand, and the evolution of water technologies, the author states, "...one tantalizing, emerging trend in the relatively water-wealthy, industrial democracies—an unprecedented, sharp productivity gain in the use of existing freshwater supplies." He goes on to describe "huge opportunities to increase effective total water supply" through the use of incentives, high-efficiency practices and technologies.

Internet Research Tips

Plunkett's Next Boom video for Chapter One:
www.plunkettresearch.com/NextBoom/Videos

Vital statistics, global: www.allcountries.org

Global population growth: see *World Population to 2300*, a 2004 publication of the United Nations,
www.un.org/esa/population/publications/longrange2/WorldPop2300final.pdf

World Population and Vital Statistics: See this extremely user-friendly database hosted by the United Nations Population Division,
http://esa.un.org/unpp

Population Clocks, U.S. Bureau of the Census,
www.census.gov/main/www/popclock.html

U.S. population projections, 2008 release, U.S. Bureau of the Census,
www.census.gov/population/www/projections/summarytables.html

International Population Database, U.S. Bureau of the Census,
www.census.gov/ipc/www/idb/worldpopinfo.php

Join in the discussion!

- See the Reading Group Guide in the back of the book.
- Go to Facebook, search for The Next Boom.
- Join The Next Boom group on LinkedIn.

— chapter two —

CONSUMERS AND CUSTOMERS; SAVERS AND SPENDERS — WHAT THEY WANT AND HOW THEY ARE CHANGING THE WORLD AROUND YOU

"Forecasting is the art of saying what will happen, and then explaining why it didn't."

-Anonymous

In the spring of 2009, I had breakfast with an old and trusted friend in a Houston, Texas coffee shop. We had our usual frank discussion about families, careers and the world in general. This man is a bright, well-educated person in his mid-fifties. He and his wife are raising a child, thinking about that child's future needs and education, dealing with issues regarding aging parents, and generally coping with the challenges and concerns that many of us have at one time or another. However, my friend and his wife have considerable advantages over many other families: they both have lucrative jobs, and they have been relatively prudent in a financial sense. Consequently, I was startled when he said, "We've stopped spending money. We've quit shopping." This, from a man who didn't shop much in the first place. He went on to relate how spending money just didn't seem like the thing to do. It might be nice to have a new car, but he recently realized he could easily run his SUV to 200,000 miles before buying a new one. They were planning a summer vacation, but were doing this at a relatively low cost since he had purchased seven nights in a luxury Caribbean resort for less than $800 in an auction on luxurylink.com, an online seller of travel bargains, including excess hotel inventory. Was this man changing his spending habits temporarily due to the Great Recession? Or is he the leading edge of the new norm? How can a conservative trend in consumer behavior today lead to booming growth in the future?

In many ways, the Next Boom is going to be on much firmer footing than the last one, because sound, long-term growth cannot be sustained by consumers who take on unreasonable levels of debt while driving personal consumption higher and higher, which is what happened in much of the 1990s and 2000s in the U.S. and Europe. Anyone who has ever tried to balance a family budget will agree. On the other hand, sustainable economic growth can be built upon an expanding consumer base of well-budgeted households that enjoy relatively strong balance sheets.

The Post-Excessive Consumption Era

To understand where we're going, you need to have a firm grasp of where we've been and how dramatically we've turned away from former bad habits. Consequently, in this chapter, I am going to focus on the retail industry to a large extent, in order to illustrate shifts in consumers' attitudes and behavior. For several years prior to 2007, American consumers had confidence that over-shopping was okay, and that buying a costly car or piece of jewelry was fun and satisfying. They could brag about how expensive something was without feeling guilty about it, even when they went into debt to make the purchase. Consumers' thinking went something like this: the money would be there; the stock market would go up; they would find a better-paying job or flip a house and make a small fortune. Home values and retail sales boomed in the glow of strong demand, while debts soared. Today, the party is over. It may not "feel" right to shop excessively, even to those who have plenty of money. Today, consumers might brag about a bargain, about how much of a discount they got, how much wear they are getting out of an item without replacing it, or how they resisted shopping altogether.

Consider the following details about retail sales and personal savings in America. To begin with, you will need to understand the concept of "same-store sales" or "comparable" store sales. This is a method of tracking the latest revenues for a given period, in a set of stores that were open during the same period of the previous year. For example, if a retail chain had 110 stores in business during the first quarter of 2009, but only 100 of them were open during the same quarter of 2008, then the percentage of change in sales is calculated only for those 100 long-term stores. In order to be truly comparable, all stores in the calculation must have been open for the entire period.

While the Great Recession was roaring ahead, same-store sales at Neiman Marcus were down 3.4% in the February through April 2008 period, and down 1.8% from May through July 2008. This was miserable enough, but a remarkable thing happened starting in the late summer of 2008: Same-store sales fell by 15.8% in August through October, 25.0% in November 2008 through January 2009 (including the critical Christmas shopping season) and 27.1% in February through April 2009. This was more than a downturn; this was a disaster from the retailer's point of view, and a sharp turn in consumers' habits. At a more middle-of-the-road chain, Macy's, which is America's largest traditional department store chain, same-store sales were off 4.6% for the 2008 fiscal year (February 2008 through January 2009). Their Christmas holiday shopping season suffered a 13.3% decrease in November 2008 and a 4.0% decrease during December 2008 (despite very aggressive sale pricing). After that awful performance, same-store sales declined another 8.5% to 9.2% in each of March, April, May and June 2009, and were down again during the critical Christmas 2009 shopping season.

There's more: Vacation spending plummeted as resorts and airlines suffered miserably during 2008 and much of 2009. Credit card use showed interesting patterns as well. Industry-leader Visa announced that, for the quarter ending December 31, 2008, debit card spending in the United States surpassed credit card volume for the first time in the company's history.[1] Visa's results show that consumers are using debit cards to a growing degree when buying non-discretionary items like food, drugs and gasoline. For some consumers, debit cards may simply be more convenient than using cash or checks. For others, however, debit cards are a way to force themselves to use available funds rather than incur debt on credit cards. "In today's challenging economic environment, debit is an ideal way for consumers to access their own funds and focus on smarter spending, budgeting and control," noted Bill Sheedy, President of Visa, Inc. for North America, in an official company announcement. Sales at many store chains showed improvement through the fall of 2010 as stock markets were up and a certain amount of consumer confidence returned. Neiman's, for example, enjoyed an infinitesimal 0.1% growth in same-store sales for its fiscal year that ended in July 2010, while it narrowed its annual loss compared to the previous year. The tide was turning for the better at Macy's as well, as same-store sales were up 5.0% for the first eight months of 2010, but the comparison represents modest growth over a miserable 2009. Retailers have two choices: adapt to consumers' new attitudes or go out of business.

Second Homes Are Out; Second Jobs Are In

Consumers are seeking financial stability—they want stable assets, fewer debts and more reliable income. They have too many friends, neighbors and relatives who are, or recently have been, out of work, foreclosed out of their homes, bankrupt or all of the above. The 2008-2010 period was scary; some people were downright terrified. Outside of Asia, sales of luxury items slumped. Consultants at Bain & Co., who follow the luxury market closely, estimated that the global market for luxury goods shrank by 8% in 2009. The decline was particularly strong in the U.S., especially among what are known as "aspirational" shoppers—people who can't afford to live a luxury lifestyle, but who occasionally splurge on luxury goods nonetheless. (Without today's growing number of *nouveau riche* in Asia, the luxury goods business would be abysmal.) American Express saw average monthly spending per customer drop dramatically from mid-2008 through mid-2009, including a 16% drop in the second quarter of 2009 compared to the same period of 2008.[2] The firm laid off thousands of people and slashed spending of all types. By September 2009, American Express was running full page ads in *The New York Times* encouraging readers to use an American Express card, "Because you pay it off in full each month, it...helps you spend responsibly."[3] American Express cardholder spending bounced back from low 2009 results by 14% for the first nine months of 2010, but the firm reports that the largest increases came from spending by businesses.

The Federal Reserve Bank (the "Fed") reported that total "revolving" debt of American households, including credit card balances, declined at a remarkable 8.9% annualized rate during the first quarter of 2009. By the year's end, the Fed found that revolving credit had plummeted 9.6% for all of 2009. The demise of flagrant shopping continued in 2010, with revolving debt declining 8.5% in the first quarter, and 7.2% in the second quarter. Obviously, the reset of consumer habits was continuing full force. The *Nilson Report*, a respected credit industry newsletter, reported that the number of major credit cards held in the U.S. fell by 32 million in 2009, an 11% drop.[4]

Some consumers proactively cancelled their credit cards. Others found themselves unceremoniously cut off, or their lines of credit downgraded, by credit card issuers. Granted, a significant portion of America's decline in total consumer debt is the result of banks writing off balances that they categorize as uncollectable, including mortgages, credit card balances and other types of loans. At the same time, however, consumers have been buying less—fewer cars and fewer houses, as well as fewer luxuries big and small. American consumers are "deleveraging" on both a voluntary and an involuntary basis.

These changes paint a picture that extends beyond the Great Recession and the high unemployment of 2008-2010. If consumers aren't spending and charging as much, then what are they doing with their money? A U.S. Bureau of Economic Analysis (BEA) report showed that the U.S. savings rate (that is, savings as a percent of after-tax disposable income) climbed from 0.6% for all of 2007, to 1.8% for 2008, further growing to 4.2% for all of 2009 and 5.8% for the first half of 2010. A trend of increased savings is being established that will likely last a long time. This recently increased savings rate may sound impressive, but it pales when compared to the 1980s average of 9.05%. (The 1990s savings rate averaged 5.83% for the decade, declining from 7.0% in 1990 to a meager 1.0% in 1999 as full-speed-ahead shoppers sailed into a sea of fiscal imprudence.)

Cash = Comfort; Debt = Sleepless Nights

American household debt as a percent of disposable income soared from 67% in tame 1980 to 133% in bubbly 2007. For the nation's consumers as a whole, incurring such an unreasonable debt load is now a thing of the past. In May 2009, a senior economist (Kevin J. Lansing) and a Group VP (Reuven Glick) at the Federal Reserve Bank of San Francisco noted that this debt ratio had already decreased to about 130%, and said "households may need to undergo a prolonged period of deleveraging."[5] In their paper *U.S. Household Deleveraging and Future Consumption Growth*, they describe a possible scenario where household debt could decline further to a ratio of 100% and the personal savings rate could increase to as much as 10% by the end of 2018. In an April 12, 2010 editorial in *The Wall Street Journal*, researcher, professor and former Clinton-era Secretary of Labor Robert Reich wrote, "Most economic analysts think a sustainable debt load is around 100% of disposable income—assuming a normal level of employment and normal access to credit."[6] By the final quarter of 2009, this debt ratio had fallen further to 122%. If debt is to return to a sustainable norm, it still has a long way to fall. Whether it is by a concerted effort to pay down loan balances (46% of households reported in a 2010 American Express survey that reducing debt was their financial focus) or by real estate foreclosure or personal bankruptcy, American consumers are clearly getting out of debt.[7]

Owning and operating a car is expensive. In America, which was historically the world's largest car market until China recently surpassed it, automobile and light truck sales sank like a torpedoed ship, from an all-time high annual rate of 17.5 million units in 2005 to 10.4 million in 2009. That's a depressing decline of 40.5%. When you do the math on these sinking sales, it's easy to see why the American automobile industry has been in so much trouble. (It took the "cash for clunkers" stimulus program to lure buyers into purchasing about 690,000 of those cars in 2009. Otherwise, the numbers would have been even worse. Meanwhile, sales in China soared 46% in 2009, to 13.6 million units.) Some consumers have figured out that they really don't need to own a car, and those who do own one have determined that today's high quality vehicles can last for years and years. Meanwhile, public transit utilization is up. "Shared" use of cars at Zipcar and other innovative services is also climbing.

Welcome to the more frugal world of tomorrow. I'm not saying that the drastic declines in spending and consumption of late 2008 and 2009 will continue long term. Instead, these numbers provide a bit of quantitative evidence of what will become a fundamental change in consumer behavior. Let's call it a first step toward the more conservative future. First, consumers were afraid and confused—they went to ground and took their money with them. The second step will be a settling in to greatly modified consumer habits and expectations. By mid-2010, consumers had moved ahead slightly, from a terrified state to a condition of worrying while spending conservatively. (That is, they were spending a bit more, but both consumers and businesses continued to suffer miserably from a lack of confidence.) The third step will be a long-term trend of much more careful household spending, fewer debts, relatively conservative saving and planning for the needs of old age. Political observer Peggy Noonan put it well, "The crash gave everyone a diminished sense of their own margin for error."[8] Higher personal savings and more frugal spending are steps that will make consumers feel a lot better for the long haul. Eventually, aging Baby Boomers who are no longer able or willing to work will draw down savings and deplete assets to fund their day-to-day needs in retirement, but the working age population is likely to be more frugal and to save at a higher rate. (See Chapter Nine for further thoughts on consumers and workers grouped by generations.)

Why You Might Consider Being an Optimist

The most recent growth spurts (the second and third legs of the 25-year-long Great Boom) ran almost uninterrupted in America from April 1991 through late 2007, according to statistics produced by the National Bureau of Economic Research. The major exception during this period was a brief (eight month) recession from March to November 2001, thanks to the crash of dotcom and technology markets, and the terrorist attacks of September 11, 2001. After that short recession, America made a remarkable and powerful recovery. That recovery was ignited to a small extent by federal stimulus spending but primarily by massive financial liquidity provided by the Federal Reserve, including extraordinarily low interest rates that remained in effect for years. Loans of all types were exceptionally, in some cases absurdly, easy to obtain, and low interest rates spurred borrowing.

Now, there are several promising aspects to changes in consumer attitudes, and these new attitudes regarding work, savings and investment are going to have a powerful effect on the near future. Consumers have learned the hard way that their former levels of savings were woefully inadequate. Homeowners have learned that they cannot count on their home equities to always rise or to be adequate for future retirement needs. Participants in retirement plans like 401(k)s have learned that stock investment values can plunge in a heartbeat. In America, the result has been a nearly instant swing toward the financially conservative side.

Higher personal savings mean lower consumption—bad news for

makers of certain goods (like high-end jewelry), retailers of non-essential items (like the now-bankrupt Sharper Image and Circuit City stores) and providers of non-essential services such as vacations, massages, pet-sitting, facelifts, personal training at the gym and gambling sprees in Las Vegas. On the other hand, increased savings combined with a growing population is good news for the long-term health of the economy, as a significant amount of money is moving away from frivolous spending and into needed investment. More money will be sitting in bank accounts, available for banks to lend. More money will go into stocks, bonds and other investments, enabling companies to conduct research and development, create jobs, build factories and purchase technologically advanced equipment. As *The New York Times* columnist David Brooks succinctly stated one view of the future in June 2009, "The American economy will have to transition from an economy based on consumption and imports to an economy with a greater balance of business investment and production."[9] This shift will aid GDP growth during the Next Boom.

How America's Economy Stacks Up

At this point in our discussion of consumers, it would be useful to draw some clear comparisons between the size of America's economy and those of other major nations, along with the proportions in which national income is spent or invested. The importance of the U.S. as a market cannot be overstated. Despite the Great Recession and the recent, often stunning, economic growth of China, India and Brazil, America still accounts for far more than 20% of global GDP. In other words, more than one out of every five dollars of the world's annual economic activity is generated in the United States. America's economy is more than ten times as large as India's or Russia's, and about three times as big as China's. America is still the Golden Land for exporters, the great marketplace across the ocean where foreign businesses long to sell their goods and services.

What Is GDP?

For those of you who are not accustomed to thinking in terms of gross domestic product, "GDP," let's step back for a moment. You probably understand that GDP is the total annual value of a nation's economy. In essence, that includes all goods and services consumed

2009 Real GDP (Est.) Selected Top Economies, at Official Currency Exchange Rates
(In trillions of U.S. dollars)

Economy	GDP
World	$58.07
U.S.	$14.26
Japan	$5.10
China	$4.81
Germany	$3.27
France	$2.66
U.K.	$2.22
Brazil	$1.49
Russia	$1.23
India	$1.09

Note: China likely moved up to 2nd place in 2010.

Source: CIA *The World Factbook*

within the nation, all government activities and all business investment. The actual calculation is much more complicated than this brief explanation. The net effects of imports, exports, inventories and other factors are also involved. GDP is an extremely useful basic measure of the economy, and it is a good benchmark for comparison of important numbers. For example: what is the level of national debt as a percent of GDP? Or, what is a nation's annual expenditure on education as a percent of GDP? Using this benchmark enables analysts to properly compare one nation to another, and to measure the performance of a nation's government and economy from year to year. A prosperous country is one that has both a GDP that grows at a rate well above the rate of inflation and a high average GDP per person.

When you are considering the future, it is important to bear in mind that government spending is a component of GDP. Unfortunately, for the prospects of the Next Boom, a significant portion of America's GDP has shifted from private sector activities to government spending. In other words, thanks to extraordinary recent growth in the size, scope and commitments of government, a massive amount of the economy and its gross domestic product involve money spent by government—not by consumers and businesses. Federal spending and debt are at the highest levels, as a percent of the economy, since the emergency efforts of World War II. In early 2010, the Congressional Budget Office estimated that America's federal government outlays would rise to 24.1% of GDP for 2010. This was soaring growth from the recent low of a bit more than 18% in 2001. Whether this high level of money for federal purposes is borrowed or raised through taxation, it is money that is rerouted from private investment to government use. Growth in state and local spending was also astounding for several years, until the Great Recession put the brakes on.

The percentage of America's national income that will be routed to taxes in the near future is an unknown—this is a potential drag on future growth of the economy. However, while government spending and debt have been rocketing ahead in America, personal consumption is heading back to more reasonable levels. If taxes are not increased significantly, then much of consumers' reduced consumption can be funneled into savings or debt reduction. In other words, a few percentage points of GDP may be rerouted from conspicuous consumer consumption, such as McMansions stuffed with flat screen TVs, into savings that can fuel investment. This can be extremely positive for the Next Boom if you take the long view. Individual debt loads may become much lower, as a percent of personal income. Investment accounts may become fuller. Future finances can be sounder and growth more sustainable. If America evolves toward an economy with less reliance on consumer purchases, it will become more like the world's other major economies.

The EU's *European Economic Forecast—Autumn 2009* provided an analysis of finances, employment, debt and growth prospects for nations in the EU, along with a comparison to many of the world's major economies. This report found that Germany relied on private consumption for only 56.5% of GDP in 2008, while that number was 57.1% in France. The number in the debt-burdened U.K. was 64.1%, and in the U.S. it was 70.1%. Despite recent declines, America's consumption rate remains extreme on the high end compared to other nations. For more perspective on this, consider recent data on China, which is extreme in the opposite direction. For 2008, China's private consumption rate was only about 35% of GDP, one-half that of the United States. This enabled China to enjoy an investment rate equal to 43% of GDP, and a household savings rate equal to about 27% of GDP.[10] For the long term, China's challenge is to get consumers to spend somewhat more, thus achieving a more balanced economy while building domestic markets. For now, however, this extremely conservative consumer mindset has helped to fund the investment and infrastructure needed to create China's current export-driven economy.

What Consumers Want and How Successful Businesses Will Reposition to Serve Them

Businesses are scrambling to reposition themselves as providers of high-value, reasonably priced merchandise. Neiman Marcus is demanding moderately priced lines from designers. In a different segment, household product makers are emphasizing lower-priced soaps and detergents, or high-value larger packages. Even companies that were already known for reasonably priced goods are repositioning. Ann Taylor, a national U.S. chain of moderately priced women's fashion stores, hired a new designer during the Great Recession and added a selection of trendier, fashion-forward clothes at reasonable prices. Thus, they have been able to keep existing customers while attracting new shoppers who want chic clothing that fits within their restrained budgets. Competitor Talbots, Inc. has used the same strategy. This is a good example of adapting to the new retail era, since many fashion-conscious women have become much more conservative about the amount they are willing to spend on clothing. "Shop your own closet first," is the new mantra of many American women who realize they can get more use from the fashions that they already own. Personal spending has shifted more toward goods and services offering quality, durability, affordability and lasting value, with less focus on the purchase of trendy items for fashion's sake. Going forward, consumers will spend their money more wisely while using debt more carefully. Successful manufacturers, home builders, services providers and retailers will respond quickly to this trend.

Whether you are observing the world of business as a knowledgeable consumer, seeking a job, making investments, running a government office or a nonprofit agency, or plotting a business strategy, an understanding of the evolving consumer mindset and how the best-run retail companies are responding can be of benefit to your endeavors, and many of these principles can be adapted to fit non-retail enterprises. When consumers spend, now, more than ever, they want confidence that they are using their money in a smart way.

Plunkett's Four Keys to Successful Consumer Products

- *High Perceived Value*: The product must convincingly offer a high level of value and durability for the price, and give consumers confidence that their money is well and wisely spent.
- *Quality and Utility as well as Fashion:* Fashion will remain important, but quality will come first in the minds of many consumers. Products that offer quality, utility AND fashion will have tremendous competitive advantage over products that offer fashion alone.
- *High Brand Reputation above Style:* The brand must stand for a company that clearly puts customer satisfaction and high value above all else. If the brand also stands for a firm with great styling, high social values, such as eco-consciousness, or other ancillary attributes, that's even better.
- *Cheap Chic Still Has a Place:* If a company wants to win the hearts of fashion-conscious, budget-conscious consumers, it must provide perky style at a moderate price—for example, cars like the Mini and the Smart. If an entire business model is based on trendy merchandise with a short useful life, then the company must make it cheaper than ever—for example, the very affordable fashions of such retailers as Sweden's H&M, Spain's Inditex and Japan's Uniqlo—a company that has been so successful at selling bargain fashions that its founder is Japan's wealthiest business person.

Perfect examples: Apple's iPod and iPhone

- ✓ High perceived value at reasonable prices
- ✓ Quality, utility and style
- ✓ High brand reputation
- ✓ Absolutely chic

Next, let's look at how these values can be applied successfully to retail stores.

Plunkett's Four Keys to Successful Retailing

- *A High Value-High Quality Product Selection:* Depth of selection is less important than a reasonably sized offering of products that the merchandiser has chosen because they consistently offer high value and quality.
- *Very Competitive Prices*: The goal here is to give the consumer confidence that the store faithfully delivers everyday low prices—meanwhile, managing the firm so as to allow the owners a viable profit margin.
- *Superior Service*: In-store help, follow up service, problem-solving, installation and repairs offered easily and quickly—the ability to make returns and exchanges must be part of the package, with an absolute minimum of inconvenience to the consumer.
- *Seamless Integration of Bricks and Clicks*: Successful firms will integrate their online endeavors with their physical presence in a manner that will provide the highest possible level of convenience to customers.

Great example: Costco

✓ Reasonable product selection, including quality store brands as well as name brands that have good reputations. However, Costco succeeds by carrying a vastly smaller merchandise selection than competitor Wal-Mart.

✓ Consistent, everyday low prices.

✓ An easy-to-find, always-staffed customer service desk. Also, rules about returns are generous and clear-cut, "We guarantee your satisfaction on every product we sell with a full refund. The following must be returned within 90 days of purchase for a refund: televisions, projectors, computers, cameras, camcorders, iPod / MP3 players and cellular phones."

✓ An easy-to-use web site with in-depth customer service information. When desired, customers may order merchandise online but return it to a store; large items, upon request, can be picked up at the customer's home for return.

Consumers Steer a New Course

After cruising through years of rarely interrupted prosperity, consumers have undergone an abrupt change from a life of shopping, borrowing and job-hopping to a more conservative lifestyle of saving, paying down debt and focusing on the job at hand. The biggest single cause of this, of course, was the Great Recession, the big crash, that sucking sound that nearly everyone heard as a big chunk of their wealth went down the drain. However, even if the crash had not occurred in such a violent and abrupt manner—if, for example, the last boom had the air let out of it slowly, at least some change in lifestyle would have come about nonetheless. I say that because the enormous mountain of debt owed by consumers had grown about as high as it could under current conditions. House prices and mortgages as a percent of household income peaked at historic, unsustainable, unpayable levels. A housing market correction was way past due. Credit card debt levels were exorbitant. Far too many consumers had fallen into a deep debtor's abyss of too much debt while acquiring items that that they didn't really need. In much of the world, from Spain to California, low interest rates, lax lending standards and overly optimistic speculation had run too far for too long.

Chastened consumers can be found in many corners of the world. Unemployment has been running painfully high in much of Europe, averaging 10% in mid-2010, a vast increase from the 6.7% of mid-2008. In Japan, businesses and workers felt the blunt end of the global recession. The Japanese auto and electronics industries were hit hard. In Dubai, planning on vast new luxury hotels and housing projects came to a halt, and many consumers who had purchased condos and BMWs during the boom in that nation found themselves out of work and unable to pay their debts. Throughout much of the world, attitudes changed considerably. Consumers and governments observed the effects of the meltdown while industry felt a devastating drop in trade, export and finance during the crash. This set of events, so well publicized, so widely felt around the world, so closely watched, worried over and reported on by media of all types, will shape the perceptions of today's younger generations and alter the behavior of older generations.

Where Companies Should Look for New Customers

So, if populations are shrinking in Japan and in many major European markets, and today's households in the U.S. will be shopping less, you might well ask, "where can businesses that sell to consumers look for additional avenues for growth?" In the U.S., part of the solution is to sell to new households formed by immigrants as well as households soon to be established by the 91 million young people who comprise America's Generation Y, those born from 1981 to 2002. Another answer is that personal consumption in China and India has nowhere to go but up, and Brazil will enjoy a very robust economic future with a growing consumer base. The same is true throughout most of the Asia-Pacific region and much of Latin America. Meanwhile, Africa is being recognized as the new frontier of retail growth, as evidenced by Wal-Mart's 2010 effort to buy a 290-store retail chain based in South Africa. Many emerging nations will evolve into terrific consumer markets over the long term. As personal incomes, education levels and infrastructure continue to improve in nations such as India and China, households will begin to consume a great deal more. Per capita income still has a long way to go in developing countries. However, both the economies and the personal income levels of India, China and other emerging nations have been enjoying substantial long-term growth, and they will continue to do so, helping to establish major legs under the Next Boom.

It can be frustrating for foreign firms to attempt to sell goods and services in China and India: bureaucracies can be extremely difficult to deal with; copyright, trademark and patent infringement are big problems in China; local regulations restricting foreign ownership of retailing, banking and other service firms can be very vexing in India. Nonetheless, nations such as these have vast potential over the long run as markets for goods and services from Europe, North America, Japan, Australia, South America and elsewhere. In the summer of 2009, for example, ExxonMobil announced a $41 billion contract for the delivery of liquefied natural gas (LNG) from its Gorgon project in Australia to China. Products as diverse as Howard Johnson hotels, Hewlett Packard computer equipment and Louis Vuitton luggage are enjoying immense success in China, while American casino firms are booming on the nearby Chinese island of Macau. (The growing global middle class is discussed in depth in the next section of this book.)

Eventually, increased household savings in America will mean that more financial resources will be available domestically, and both purchases and borrowing from nations like China will be lower than they would without the savings trend. Meanwhile, in China in particular, as well as Japan, Germany and other countries where a vast portion of the economy is geared to export, the challenge will be for those economies to wean themselves from sales to America, by boosting domestic demand and household consumption, and by growing exports to the developing world. India is less export-driven, and India's sales to the U.S. are more likely to be business services than consumer items. The booming middle classes in Brazil, India and China will create growing domestic consumption and thriving local businesses. Nonetheless, selling to America will remain important to all nations, due to the massive size of the U.S. economy. Buying from America will remain important as well, as long as American companies maintain the competitive advantages they currently hold in many categories of technologies, goods and services. Meanwhile, the nature of this trade is evolving. Emerging nations will be strong markets for exports of technologies, commodities, products and expertise from North America, South America, Australia, Europe and Japan. Economists at JP Morgan Chase estimated that emerging markets would represent 34% of global consumption in 2010, compared to only 23% twenty years ago.[11]

LOHAS: Socially Conscious Consumers Create Challenges and Opportunities for Advertisers and Marketers

An additional consumer trend that will play an important role in shaping the near future is LOHAS, an acronym for Lifestyles of Health and Sustainability. This is a term used to describe that segment of consumers whose purchases are influenced by matters such as corporate social responsibility, recyclable materials, energy efficiency, organic contents, toxicity, allergens, environmental impact and alternative living styles. "Eco-friendly" products are important to them, but these consumers should not be confused with extreme "greens" or environmental fanatics. LOHAS consumers often prefer to buy organic or "natural" foods, dietary supplements and personal care products; they also often prefer alternative medicines and therapies, in the form of acupuncture, massage or herbal remedies. Furthermore, these consumers tend to be strong advocates for renewable energy and they may seek out products from socially conscious companies.

LOHAS consumers are from all age groups and income levels. Although this group of people is far from homogeneous, it represents a significant portion of the consumer market. Their enthusiasm for LOHAS products, services, stores and brands ranges from minor interest to dominance of their purchasing decisions. Some of their issues have begun to gain popular momentum. In the face of growing levels of obesity, lapses in food safety and the world's massive fossil fuel consumption, consumers' concerns about issues such as healthier foods, workplace standards and renewable energy are having a strong effect on their shopping practices.

As a result, many more companies have begun to present themselves as environmentally and socially conscious entities. Many have taken steps to build greener facilities, or to place stricter controls on the working conditions at factories run by suppliers. Many also contribute significant sums to nonprofit organizations that support LOHAS issues. One recent example of this corporate trend is Gap, Inc., which in 2004 released its first-ever "Social Responsibility Report." In an unusual move, the clothing retailer admitted to problems with some suppliers regarding work and safety conditions, and revoked its approval of 136 factories. In the global business arena, social responsibility is an important issue that many companies must address. Meanwhile, the green business trend is showing up in virtually all sectors. For example, resorts worldwide are anxious to bill themselves as leaders in ecotourism. Recently, American consumer magazines as diverse as *Shape* and *Town and Country* devoted entire issues to environmentally friendly topics. Advertisers jumped to provide eco-sensitive ads highlighting goods that fall into green categories, such as those using recycled packaging or non-animal product testing.

Recent trends in high energy costs have combined with the global financial crisis to boost the LOHAS mentality and create interesting offshoots. Since consumers are now much more financially conservative, and they want items that are of lasting value, a "less is more" mentality could spread. For example, the continual growth in the size of the average new home built in the U.S. has stopped, and homes will be smaller, but smarter, going forward. The same is true in automobiles, as evidenced by the tremendous success of the small, but smart, cars like the Prius and the Mini. Health care costs and other considerations are boosting demand for products that promote a healthy lifestyle.

Savvy Retailers, Manufacturers and Marketers Will Adopt the Following Practices in Order to Position Themselves for the LOHAS Market

- Stress an organization's sensitivity to environmental issues, energy concerns, personal health needs and restrained personal budgets.
- Be aware that many consumers will pay more for LOHAS-centric goods and services, but only when they see lasting value or reduced environmental impact. For example, millions of consumers have paid higher prices for hybrid-equipped vehicles than they would have paid for traditional cars, despite the fact that it can take many years to earn a return on that extra cost through reduced gasoline use.
- Day-to-day consumer products must be priced within reason, even if they have high LOHAS factors.

Perfect examples:

✓ Whole Foods (they are successfully defeating what was once a "whole paycheck" perception of their pricing while advancing their image as a LOHAS-based firm)
✓ Trader Joe's (they benefit from an everyday low prices image, along with an air of sustainability and social consciousness)
✓ Toyota Prius (fuel efficiency)
✓ Zipcar (why buy a car when you can call for one on an as-needed basis?)
✓ NetFlix (no need to drive to the video rental store, plus reusable packaging)

Internet Research Tips

Plunkett's Next Boom video for Chapter Two:
www.plunkettresearch.com/NextBoom/Videos

The Federal Reserve Bank publishes a wealth of economic data that is easy to access and easy to use at www.federalreserve.gov

The 12 regional Federal Bank branches publish extremely interesting data of their own. Among their web sites, my favorite is the exceptional site operated by the Federal Reserve Bank of Minneapolis, www.minneapolisfed.org

The CIA World Factbook publishes easy to use data about the economies of the world's nations at:
www.cia.gov/library/publications/the-world-factbook/index.html

The Bureau of Economic Analysis, www.bea.gov offers easy-to-use and customizable tables on all aspects of the U.S. economy.

Join in the discussion!

- See the Reading Group Guide in the back of the book.
- Go to Facebook, search for The Next Boom.
- Join The Next Boom group on LinkedIn.

— part two —

PLACES

— the next boom —

— chapter three —

GLOBAL TRADE 2.0 AND THE PHONE LADIES OF BANGLADESH

"My interest is in the future because I am going to spend the rest of my life there."

-Charles Franklin Kettering (American inventor, 1876-1958, head of research for General Motors)

Much of the impetus for the Next Boom will come from emerging nations—those countries around the globe that will continue to enjoy soaring economic growth as their levels of commerce, education and infrastructure escalate to new heights. Consequently, Part Two of this book focuses on important factors that will combine to create unprecedented opportunities for business and investment: the world's quickly expanding middle class, an extension of modern services and technologies into developing nations, and the coming evolution in global trade.

A Day in New Delhi—Not Walking

I generally walk miles and miles in every new city I visit—sometimes spending an entire day tramping 10 or 12 miles through neighborhoods to get the local flavor. On a recent trip to India, my habit came to an abrupt halt in New Delhi. All I could do was walk around the block once, because to me, as a westerner accustomed to a reasonable level of protection as a pedestrian, it was nearly impossible to cross the street. There are no crosswalks and few traffic lights in central New Delhi. Traffic uses rotaries or roundabouts—I suggest that you don't try being a pedestrian in a six-lane roundabout jammed with cars, trucks, scooters, buses and the occasional elephant. If I could magically be leader of India for a day, I would issue a decree calling for the construction of 1 million pedestrian bridges. (The city of Mumbai is now experimenting with a small network of pedestrian bridges.)

Leaving Delhi by car for the adjacent high-tech suburb of Gurgaon, we crawled through traffic, even on the broad new multilane freeway that was intended to efficiently support Gurgaon-bound traffic. Hundreds of unfortunate pedestrians gathered on each side of the freeway were hoping to dash across a dozen lanes of traffic so they could continue on their way on foot. This would be similar to attempting to walk across a Los Angeles freeway at rush hour.

In Gurgaon, a relatively new development of skyscrapers filled with entrepreneurs, analysts and software engineers, modern civilization is advancing in fits and starts. Electric lines dangling from poles on some street corners look like massive tangles of spaghetti instead of orderly wiring. A hog nosed about for food on an empty lot next to a modern multi-story building. Nearby, a cow stood chest high in a garbage dumpster, foraging for supper. The stark contrast between proper construction and barely supported structures, good highways and impossible traffic, modern infrastructure and barely-on electricity, even in this new city, the Indian equivalent of Silicon Valley or the Boston area's Route 128, is a fair representation of both India's challenges and its remarkable progress. The scene made me wonder whether India could ever catch up to its infrastructure needs. In the smaller cities and agricultural villages, a lack of modern infrastructure is much more pronounced.

This description of urban problems is not intended to take away from India's significant progress. For example, I spent half a day in the high rise headquarters of Evalueserve, a purveyor of outsourced research services that has grown at blinding speed. This is the type of Indian company you might envision, ballooning from a couple of hundred people when I first met with them in 2003, to about 2,500 today—MBAs, engineers, analysts, a mass of bright young people sitting in cubicles analyzing market shares, trade patterns, and industry's needs around the world for blue ribbon clients. Evalueserve was started by a small group of highly educated people who were formerly with IBM, or were consultants at one of the world's leading management consulting firms, McKinsey & Company. They realized that they had a unique opportunity to hire some of the thousands of eager new graduates who are pouring out of India's booming universities each year, and put them to work at modest prices serving the competitive intelligence needs of the world of business. For example, an investment firm in the U.S. might want a 20-year forecast of the rate of growth of uranium usage by the nuclear power industry worldwide. Evalueserve can promptly assemble a qualified team to prepare a detailed report on such a topic, and deliver it quickly at perhaps one-half the price that a company in the U.S. might charge. In the world of business, this industry is referred to as KPO—knowledge process outsourcing. There are hundreds of successful companies in India that specialize in this field.

Over tea in Evalueserve's main offices in Gurgaon a few months ago, COO Ashish Gupta explained to me how the company had recently opened offices in places like China and Chile, in addition to Europe, India and the U.S. It now has people available in the world's most important markets and in the most vital time zones. For instance, if a client in Mexico City wants to talk to an analyst in the Spanish language in the middle of the business day, the office in Chile can handle it. Ashish is typical of the leadership of the firm—formerly at McKinsey & Company, educated in America with an MBA from Carnegie Mellon, in addition to his engineering degree from the highly regarded Indian Institute of Technology. In the U.S., I had lunch with the firm's chairman and co-founder, Alok Aggarwal. He holds a Ph.D. in computer science from Johns Hopkins University, as well as a degree in electrical engineering from the Indian Institute of Technology. Alok was formerly the Director of Emerging Business Opportunities for IBM Research Division worldwide. These are the types of highly competent, extremely well-educated, very competitive people who are reshaping the world in the new era of global trade.

I visited beautiful, modern homes in Gurgaon that would rival similar homes in America, and offices that were perfectly modern and teeming with talented, well-educated technicians, analysts and professionals. I was invited to an elegant lunch in the home of an exceptional young couple. Both had excelled at an American Ivy League university. He has family in America, and they could have easily built a successful life together in any city in the U.S. However, they decided to live in India, her family home, where he is a respected entrepreneur and she is a high-powered business attorney. These extremely intelligent high achievers see a very bright future for themselves, and for their children, in the heart of urban, high-tech India rather than in America. To me, they are an indicator of the potential for India's growth and its future powerful role in global business.

Challenges for India and China

There will soon be a lot more money to be made in the developing world. Less-developed nations already represented 82% of the world's population in 2008 and will account for 86% in 2050.[1] These countries will be responsible for the vast majority of population growth between now and 2050. As household incomes rise in these nations, and they will grow dramatically, business opportunities will be immense.

"Globalization" is a somewhat overused word, and it has negative connotations to many people. Since I believe that we are going to be rocketing to new levels of trade that will be generally positive to the global economy, I will refer to the enhanced world of business that will evolve over the near future as Global Trade 2.0 ("GT2"). However, GT2 will be full of surprises. It will not always look like what you might expect it to.

Are China and India economic miracles that will continue to post incredible growth year after year? Growth will eventually moderate, but trends point to decades of significant expansion in their GDPs. China and India today are somewhat analogous to America of the late 1800s—a booming young nation that had resources in place to start modernizing and building the economy at a very high rate, a process that continued with a few interruptions until the Great Depression of the 1930s. For example, railroad mileage in the U.S. soared from 30,000 miles in 1860, to 163,000 miles by 1890, and 248,000 in 1920,[2] allowing for very rapid industrialization. The establishment of telegraph and telephone service, electric supply, roads and water works expanded at a high rate, further enabling business and industry to grow. Similar patterns are occurring now in India and China, with modern infrastructure growth that includes cellphone towers, highways, shipping ports, airports, waterworks, electric plants and railroads. (In the recent past, we also saw this pattern of growth in South Korea, Singapore, Taiwan and Japan.)

Although there were some problems, India and China barely stumbled during the Great Recession. India's banks had been conservatively managed, and weren't forced through the gut-wrenching pain that enveloped the banks of the Western World. In China, the government promptly poured money into public infrastructure construction projects during the recession, while opening the spigots at Chinese banks to flood industry with loans and fund consumer purchases of cars and homes. It remains to be seen whether these loans will eventually create bad debt problems or excess inflation. China was unusually proactive and effective in facing the recession, and its massive new investments in badly needed infrastructure were timely and reasonably effective.

From a structural point of view, China faces dramatic changes. Li Keqiang, a high ranking and closely watched Chinese official, made a speech in 2010 noting China's biggest challenges. Despite the nation's tremendous record of economic growth, he cautioned that it must, "change the old way of inefficient growth and transform the current development model that is excessively reliant on investment and exports...We will focus on boosting domestic demand."[3] He was referring to the need to evolve, to become a sustainable economy based on growing local markets, greater spending by households and rising incomes in addition to its powerful export engine. Also, curtailing pollution and environmental destruction will be a vital part of moving China's economy to the next level. In a nation of more than 1.3 billion people, these are developmental and economic tasks on a greater scale than any nation has ever faced before.

Global Trade 2.0—A Tripling of Trade by 2030 Will Raise Personal Incomes, Build the Middle Class and Create Massive Business Opportunities

How will world trade evolve over the near future? In *Global Economic Prospects 2007*, the World Bank states, "Global integration is likely to enter a new phase. In virtually every growing economy the importance of trade...will rise, continuing the trend of the past two decades. The growth...over the next 25 years will be powered by a new dynamism in global trade within [the services sector]."[4] The World Bank goes on to forecast that trade will rise threefold by 2030, to $27 trillion. Emerging countries will benefit greatly as they continue to modernize. Markets for services will boom along with rising household income, including domestic demand for entertainment, restaurants, retailing, communications, education, transportation, residential services and health care. The challenge for mature nations is to remain innovative and competitive, using their strong bases of higher education and advanced technologies to create products and services that the developing world will want to purchase in large quantities. Many U.S. and European firms, such as GE and Siemens, already have strong efforts in this regard.

Although they will certainly evolve and adapt, China, India, Indonesia, the Philippines and similar offshore work centers will remain moderate-cost, highly competitive providers of services and manufacturing for the near future. However, local costs in these countries are rising—eventually, they will rise substantially. At the same time, as their economies grow, their business structures and middle classes will grow as well, and they will offer lucrative markets for exported intellectual property, technologies, certain manufactured goods and high-level services created in the U.S., Europe, Japan and elsewhere. In particular, these emerging nations can be spectacular markets for major brands of consumer goods. While the brands (Gillette, Apple, Starbucks, Nestle, Kleenex, KFC, etc.) will be American or European, much of the manufacturing will be done in local markets such as Guangzhou, Mumbai, Jakarta and Manila.

Meanwhile, these developing nations face immense challenges prerequisite to substantial future growth, including the need to:

- Greatly enhance infrastructure such as roads, highways, railways, airports and electricity networks
- Extend and improve public education systems, particularly into the villages
- Control pollution and clean up existing environmental damage
- Provide access to basic health care services
- Create a social safety net that includes unemployment insurance, pension plans and retraining for laid off workers
- Foster a healthy level of local consumer demand while evolving, from economies highly dependent on exports, to economies with vibrant domestic markets
- Provide greater economic opportunities to residents in rural areas

- Enhance energy efficiency
- Promote property rights and rule-of-law while combating graft and corruption

Wake Up, Your Hotel May Be Under Attack!

The above list contains structural challenges. Then there are the political problems to consider. India and China combined hold about one-third of the people on the planet, many of them jammed into some of the world's largest cities. Within each nation there lies a mind-numbing array of religions and philosophies, political factions ranging from communists to anarchists to radical religionists, and languages spoken in one part of a nation that can't be understood in another, along with traditional tribes, castes and even a few nomadic groups. The potential for strife and conflict is beyond measurement, and it is often compounded by agitators from neighboring nations. This was brought home to me in a personal way while I visited New Delhi in November 2008. Despite the extra fees I was paying my U.S. cellphone provider for international use, and the firm's highly convincing ads about ease of access abroad, my handset only worked once out of the dozens of times that people tried to call me. The single call that arrived woke me up in the middle of the night at the Taj Mahal Palace hotel. "Wake up, your hotel may be under attack," someone from my office said. That statement woke me up all right. Coverage of the horrible terrorist attacks in Mumbai were the focus of worldwide news that day, and the Taj Mahal Palace was at the center of the tragedy that eventually claimed 170 lives and injured 400. Fortunately, I was at the Taj Mahal Palace in Delhi, not Mumbai—an easy mistake for someone back home to make in the panic of the day. Nonetheless, I didn't sleep much during the rest of the trip, and my hotel was soon enough surrounded by army personnel toting guns, while the lobby was awash with men in dark suits, stationed every few feet, wearing tiny earphones. A metal detector was installed at the front door, and purses and packages were searched at the entrance. Cars were inspected at the driveway gates by a swarm of guards wielding under-car mirrors on poles. (Nonetheless, a passerby could easily have lobbed a small explosive device over the wall surrounding the hotel's grounds.) All of this within a few yards of the houses of government and the presidential palace. Polite settlement of political and religious disputes is not something you can take for granted in the developing world, and this problem will remain a serious impediment to growth in far too many nations.

Falling Off the Train, Five to a Scooter

In addition, rapid growth in offshoring centers in China and India has created myriad shortages and problems. The most popular Indian business centers, such as Bangalore, Mumbai and Gurgaon, are experiencing daunting shortages of infrastructure of all types. A simple, but horrifying, example: Thousands of people die each year when they tragically stumble or are accidentally pushed off India's vastly overcrowded commuter trains on the way to work, or are hit by trains as they try to walk across the tracks—a gruesome fate that Indian commuters risk daily. Despite the recent construction of new highways in India, traffic delays and inefficiencies are an immense burden. India's traffic deaths were the highest in the world in 2008, at 118,000.[5] If you have a mental image of India's work centers as modern communities that look like Atlanta, you have the wrong idea. Instead, there are stark contrasts between the modern and old worlds, often in the same city block. In India, my car shared the road with a working elephant—lumbering along on the way to a construction site, driven by two young men, "*mahouts*," sitting on a wooden platform—in the middle of New Delhi on a major street near the halls of government, a street equivalent to Pennsylvania Avenue in Washington, D.C. At various times, in India's cities, we shared the road with bicycles, goat carts, ancient tractors serving as taxis by stacking passengers like bowling pins on wooden decks perched behind the drivers, buses groaning with riders packed into their interiors like chickens in a shipping crate, and motor scooters carrying families of five on one small seat, weaving through the traffic in a motorized ballet. For some reason, if anyone on a scooter is wearing a helmet, it is the man who is driving, while his family rides unprotected. Traffic deaths are common and horrible, and entire households are frequently wiped out in a single collision.

Services vs. Manufacturing

India is already focused to a large extent on service industries and knowledge workers. However, this nation is also home to several industrial giants, and leading Indian companies such as Tata Group and Reliance Industries are investing substantially in manufacturing businesses such as steel products, chemicals, textiles and automobiles. There is also a booming pharmaceuticals industry in India.

While India thrives on service businesses, China remains mired in an export-driven, manufacturing-intense economy. However, China has a unique opportunity to partner with its neighbor Japan in order to evolve. Despite its immense investments in universities and research parks, China continues to need to boost its innovation and intellectual output. While China could benefit from additional expertise in areas such as software, electronics, renewable energy and environmental controls, Japan has abundant intellectual resources in these fields. This will be a powerful partnership for the long haul—in 2008, China became the largest export market for Japan's products, a distinction formerly held by the U.S. Japan is also one of the largest direct investors in China. Meanwhile, Japan, with its dwindling, aging population and shrinking domestic market, needs China's abundant capacity to provide outsourced services and manufacturing for the Japanese. This could be a spectacular economic match if a proper foundation is laid. Meanwhile, some Indian firms are developing new strategies in order to gain market share in Japan. Wipro Technologies, a leading Indian outsourcing firm, saw its revenues in Japan grow from less than $40 million in its fiscal 2004 period to $115 million in 2009.[6] Indian firms are educating some of their workers in Japanese language, culture and business practices in order to promote business in Japan. This reminds me of the way in which my sons studied the Japanese language and culture, laying the groundwork for their own success in the world of business in Asia.

The growing economies of Asian nations such as China, India, Malaysia and Indonesia are neither foolproof nor recession-proof. Crashes can and will eventually happen. For example, Japan showed such exceptional growth in the late 1970s and early 1980s that many observers thought it could grow to be the world's largest economy. Technology-based manufacturing was soaring, and Japanese makers of consumer electronics and automobiles, such as Sony and Honda, had become global leaders in terms of quality and branding. Fueled by easy lending and over-investment, Japanese real estate and stock market values reached insane heights. The eventual crash was devastating. As we have seen in other fast-growing Asian nations that are now relatively mature, such as South Korea and Thailand, financial crises, currency calamities and other problems can, and do, occur. Nonetheless, in a relatively short few decades, Korea, Singapore and Taiwan, like

Japan, became true economic powerhouses with vibrant middle classes, and they matured into exceptionally successful nations despite their relatively small populations. The same level of widespread economic and social success could be developed in China and India over the long-term.

Entering the Middle Class

The future rise of vast numbers of low-income people to the middle class will create the biggest marketing opportunity the world has ever witnessed. Wal-Mart already has 137 million visitors to its U.S. stores in an average week. How large will this number be on a global basis when the firm expands ever more deeply into Latin America, Asia and eventually Africa? 500 million? 1 billion? China is already a bigger market for GM's cars than its home nation of America. Amazon.com, Avon and McDonald's would not be the companies they are today without their massive overseas customer bases, and they have barely scratched the surface of their potential growth.

Think of "middle class" as a condition that indicates, at the very least, that a family has a steady flow of a small amount of discretionary income that can be spent on luxuries or non-essential items. Solid entry into the middle class means that a household may own a scooter or a car (instead of walking, or riding a horse or a bicycle). A middle class family may own modern consumer electronics or be able to pay school tuition for its children. It is difficult for a Westerner to imagine the living conditions in a rural Asian village without seeing it firsthand, but hundreds of millions of people are living in the most basic conditions imaginable. Nonetheless, this is changing steadily. The World Bank forecasts that, by 2030, the number of people living on less than $1 per day (extreme poverty) will drop to 550 million, a decline of 50% from 1.1 billion in 2007.[7] Looked at in broader scale, they forecast that people living on less than $2 a day should fall below 1.9 billion by 2030, a reduction of 800 million from 2007 levels.

Mobile Phones and the Next Boom

Mobile phones have the unique ability to leap past barriers. No highways? No landlines? No banks? No Internet connections? No problem. Cellphones create instant links to modern civilization that can be carried around in subscribers' pockets. Low-cost, easy to erect service towers solve the logistics problem instantly. State of the art telecommunications equipment made by companies like China's Huawei provides the network systems at relatively low cost. This is true even in the deepest, darkest corners of the world, from Nigeria to New Guinea, from Cambodia to the Congo. By 2010, there were already more than 5 billion cellphone subscriptions on Planet Earth—providing instant, mobile communications to more than two-thirds of the world's total population! Even people living the simplest lives imaginable, with no running water, dirt floors in their homes and cooking over wood fires, have cellphones. Soon enough, virtually every adult on Earth who wants one will have at least one cellphone, as will a vast number of their children. In India, for example, innovative business practices have created the world's most efficient cellphone companies, bringing the wireless world to subscribers at rock bottom prices. Traveling salesmen come through villages and put on entertaining shows in the village squares, extolling the virtues of the latest handsets like traveling potion salesmen of late 19th Century America.

India's base of cellphone subscribers was growing at a rate of several million monthly in 2010. Here, service providers are extremely aggressive, and their prices are among the most competitive in the world. India's Bharti Airtel had 135 million cellphone subscribers as of mid-2010. At some firms, cellphone calls cost $0.02 (U.S.) per minute or less. In February 2010, Vodaphone announced a simple new phone, the 150, which it will sell for under $15 in Africa, which is a non-subsidized price. For less than $20, consumers can buy the model 250 which includes such upgrades as a larger color screen and an FM radio.

More Cellphones = Higher GDP

A recent study at the World Bank determined that an increase of only 10% in the number of people using cellphones in a developing nation will increase the growth of GDP in that country by nearly 1% (0.8%).[8] How does the spread of cellphones boost economic growth? Because modern communication enhances the efficiency of nearly any type of activity—especially in business and industry. A craftsman who obtains a cellphone can now advertise his trade along with his phone number. A shop owner who has a cellphone can call various wholesale suppliers to find the best prices on merchandise. Here are a few more examples of innovative uses for cellphones that boost economic activity in Asia and Africa: Farmers are able to pay very modest amounts to subscribe to text message-based agricultural information services, including weather forecasts and market prices. China Mobile runs one of these services, and there are several examples in Africa and elsewhere. In some systems, such as "Farmer's Friend" in Uganda, subscribers can send text queries about crop problems, pest control or other needs, and receive expert advice promptly via text. Other systems like Google Trader enable farmers who are ready to sell to be matched up with buyers of produce and farm animals. Rural farmers are finally able to gain prompt access to competitive market prices.

There's more: In 2008, the UN Foundation and the Vodafone Foundation identified more than 50 mobile health information services in developing nations, where data on prenatal care, drug safety, HIV and other health issues are distributed wirelessly by text messaging. In some cases, cellphone cameras are used to transmit photos of patients' conditions to remote clinics for diagnosis. The introduction of prepaid cellphone service, often purchased via "top-off" vouchers in developing nations, was revolutionary. This means that consumers, once they have purchased a handset, are not required to have good credit histories in order to get phone service—they simply need a modest amount of money with which to prepay for calls. The equivalent of fifty cents will buy half an hour of cellphone time.

One of the more desirable careers in developing nations is that of the Village Phone Lady. Grameen Bank and Telenor, a telecommunications firm based in Norway, successfully teamed up with other organizations to found GrameenPhone. Their vision was that telecommunications could play a major role in poverty reduction in Bangladesh, and that no one in a rural village should be more than a 10 minute walk from a telephone. GrameenPhone was founded with a relatively modest $125 million. It received a broad license to provide cellphone service to both urban areas and Bangladesh's 68,000 rural villages. Service was launched in 1997. Among other things, Grameen Bank provided small loans to women in remote villages. With these loans, the women purchased mobile phones with prepaid airtime so they could set up businesses that let their neighbors use the phones to make calls, for the equivalent of a few pennies each. This is a big step forward in villages where residents have never had access to any type of telephone. Eventually, there were more than 200,000 Telephone

Ladies in Bangladesh. The Telephone Ladies are able to repay their loans after a few short months.

This is part of the relatively new global concept of "microloans." Well-organized lenders make loans that typically equal a few hundred dollars to aspiring entrepreneurs. The borrowers often buy tools, such as sewing machines, that enable them to establish or grow their businesses. Without the availability of this capital, these people might be stuck in lives of poverty. Many Telephone Ladies use their newfound earnings to fund additional lines of business. Some use their profits to open shops or cafes. Others use the money to improve their family's farm. One early Phone Lady, Mosammat Anwara in the village of Chamurkham, Bangladesh, reported earning a net $80 monthly.[9] The money was transformational to her family. She was able to pay for the educations of her four children and establish a small poultry and fish farm. At the time, the average annual per capita income in Bangladesh was only $250, and generally much less in the villages. Some Phone Ladies were able to net, from their small telephone businesses, three to four times that amount.

Mobile Banking—A Revolutionary Concept

However, sustained economic progress is not possible without widespread access to a reliable consumer banking system. Consequently, the biggest single contribution that cellphones make to economic growth in the developing world may be mobile banking—enabling people in remote villages to have access to bank accounts for the first time in their lives, including the ability to safely build savings, make remote payments and send money transfers by wireless means. Companies tapping this potential goldmine include M-PESA in Kenya, which is run by Safaricom, a mobile service provider partly owned by Vodaphone. East African company Zain offers the Zap mobile money transfer service and South Africa's MTN plans to deploy MobileMoney accounts in as many as 21 African and Middle Eastern countries. mChek, based in Bangalore, is a very innovative company to watch. It has partnered with Airtel to create a mobile payment system that is rapidly gaining ground in rural India, where physical banks are hard to find, and it is also providing mobile convenience to city dwellers. mChek links customers' debit cards and credit cards to their mobile phones, enabling them to pay bills, send and receive money transfers, and make purchases for things like movie tickets. The firm uses state of the art security and encryption technology.

By 2010, more than 30 Indian banks had received approval to provide mobile banking services. In rural India, a one-man banking stand can take deposits from customers, who are now able to establish bank savings for the first time in their lives. Their accounts are credited by the banker via a code sent to a main bank by text message. The account's owner has a user name and password to enter via cellphone, enabling the customer to identify herself and track account balances. When bills need to be paid, money can be sent by the same methods to another bank customer's account. These mobile banking customers are able to purchase and pay for goods or services via long distance. Those with mobile Internet access can view and order online, purchasing, for example, a ticket on one of the developing world's many new, ultracheap airlines. Mobile consumers are also enabled to receive government information, health care assistance, news, entertainment and educational material via their cellphones. They are connected to the modern world. They can save and spend money. They can engage in trade and commerce in order to boost their own incomes. With hard work and a little luck, they can climb up the first steps of the ladder to the middle class. Their children who have left the villages to go to distant cities, in order to find work, can use cellphones to stay in touch with family members back home via text and voice messages. Even more important from an economic point of view, they can send mobile remittances back to their parents who now have mobile banking accounts.

China is already the world's largest cellphone market. China Mobile Ltd., the leading provider, crossed the one-half billion subscriber mark in the third quarter of 2009, and reached 569 million subscribers by September 2010. This nation's cellphone market is evolving rapidly, and China's Ministry of Industry and Information Technology reports that the Chinese will invest tens of billions of dollars in advanced 3G, or third-generation, cellphone networks in the near future. This equipment, which will come largely from Chinese manufacturers such as Huawei and ZTE, enables mobile Internet access at high speeds.

Cellphones lead to innovation at multiple levels. For example, consumers who have mobile phones also need a way to recharge them. In a manner reminiscent of the way that the developing world uses cellphones to leap past the problem of no landline infrastructure, many cellphone owners are utilizing innovative, renewable electric supply strategies, including solar chargers, to overcome the fact that they are off the grid.

Closer is Better, or RCT Regional Centers of Trade

While overall trade will grow during the Next Boom, the nature of that trade will evolve. Most people are well aware that the export of manufactured goods from China has been the key to that nation's astonishing growth over recent decades. This began with relatively low-value, unsophisticated goods (such as toys, household items and apparel). For example, according to the American Apparel & Footwear Association, Americans purchase more than 2.2 billion pairs of shoes yearly. More than 98% of those shoes are imported, and China is the largest supplier by far. These basic exports will remain of great importance to China. Nonetheless, China's manufacturing base slowly evolved and moved ahead so that a significant portion of its exports is now comprised of sophisticated gear such as telecommunications equipment. Such a shift in the level of manufacturing involves multiple changes. For instance, the Chinese ramped up their engineering and research to the extent that Huawei Technologies became one of the world's leading telecom equipment makers over a short period of time, because of local engineers who either adopted well-proven ideas from other nations or innovated new designs of their own. The quality is high and the prices are low. Meanwhile, former global giants in this field, including Nortel, headquartered in North America, which filed for bankruptcy in 2009, suffered miserably as they lost market share to lower-priced Huawei goods.

However, some of China's advantages have begun to slip perceptibly. The most talented, experienced Chinese workers were in short supply during the last boom. By 2010, soaring real estate prices and workers' demands for significantly higher wages were acute problems. Then there is the threat of high shipping costs for the transport, to distant markets, of goods manufactured in China. From 2001 through 2007, container shipping traffic on the Asia-Europe route grew by about 15% per year, according to Drewry Shipping Consultants Ltd. The price for shipping a large container of goods from Asia to Europe in 2007 ballooned to $2,800. Container shipment costs swing wildly, depending on supply and demand. At 2007 rates, the price of shipping goods from China had become prohibitive, absorbing profits and ruining the economics of some industries. By 2008, the Asia to

Europe price plummeted to about $700 per container, as the Great Recession put the brakes on demand and oil prices fell dramatically. Importers in North America and Europe are acutely aware that shipping costs could rocket to new heights in the future. Meanwhile, there is a growing outcry about the environmental impact of intercontinental shipping.

India has likewise endured increases in costs as demand from the booming software, consulting and call center industries drove up prices for personnel, facilities and operations. The CEO of an Indian firm explained to me over breakfast in 2006 how he had budgeted a steady 15% per year for wage hikes. This could easily turn out to be much too low. Increases of that magnitude would drown a labor-intensive company in higher-cost America. Wages in India are likely to grow at even faster rates during the Next Boom, as shortages of educated, experienced workers could become a significant problem.

The metamorphosis of industry in China and India is reminiscent of the evolution of the Japanese automobile industry after World War II. At first, Honda and its peers built all of their vehicles in Japan. Their products were simple in nature, and they sold at modest prices. The venerable Honda Civic, for example, was first sold in America in the early 1970s, a relatively fuel-efficient vehicle that was well timed to serve consumers who were facing the 1973 oil crisis and long lines at gas stations. By the 1990s, however, Japanese-owned manufacturing plants were commonplace in the southern Unites States, churning out very sophisticated, top quality cars for eager American consumers. While much of the research and engineering is done in Japan, and some of the components may be built offshore as well, the final assembly is done in America, largely for local consumption. (The same is true of many German and Korean models that sell well in the United States.) From a cost and efficiency standpoint, this is completely logical. Thanks to modern communications, there is essentially no cost to delivering the engineering and intellectual capital from Japan to the U.S. The finished cars, weighing one or two tons apiece and extremely expensive to ship long distance, are assembled in the U.S. within a short distance of the dealerships that will sell them. This gets the cars to market more quickly, and enables the manufacturers to make rapid responses to dealers' needs for inventory in certain quantities, colors and configurations.

Regional Trade Evolution

The Chinese are using their immense stores of foreign currencies to acquire companies within foreign markets, including America and Europe. Likewise, Indian consulting firms have been acquiring American subsidiaries and opening ever-larger U.S. and EU offices. Meanwhile, China has been taking the lead in fostering regional trade agreements (RTAs) throughout its part of the world. In fact, RTAs have been blossoming worldwide in recent years. The World Trade Organization (WTO) reports that it was notified of the signing of 300 new RTAs in a mere 13 years from 1995 through 2008. In contrast, only 124 such agreements were officially noted from 1948 through 1994, a period of 46 years.[10] This acceleration shows that many nations are focusing on a regional trade strategy. The trend is particularly strong in Asia. It is Global Trade 2.0 in action.

Meanwhile, in mature markets such as America, some small- to mid-size manufacturers are downsizing their offshore efforts.[11] They have found that it is simply too inefficient to attempt to manage supply chains, engineering or manufacturing centers that are thousands of miles away from their headquarters back in America, Canada or the EU. For a firm of 100 to 1,000 employees, attempting to send valued managers back and forth to Asia on a continuous basis is exhausting, expensive and inefficient. Local and regional centers of trade and manufacturing may evolve as a result. For example, the *maquilladores* manufacturing centers on the U.S.-Mexico border (where goods move from manufacturing plants in Mexico to distribution or final assembly centers nearby on the U.S. side) may get a new boost. GT2 will foster this trend, as companies from China, India, Germany, Brazil, Taiwan and Korea make investments or enter into partnerships in their core markets such as America and the EU. Brazil's JBS SA, the world's largest meat company, has acquired such American meat firms as Pilgrim's Pride. Capital for investments and expansions will flow to promising RCTs for valid projects. Just-in-time inventory will be boosted, as some components and finished goods will be produced closer at hand. Personal interaction might grow as well. For instance, services providers would find it easier to visit their most important clients if they are located in nearby RCTs.

The U.S. and Canada have tremendous potential to benefit from expanding regional trade agreements and activities with Latin America, a region that will enjoy significant growth over the near future. While the size of its population is nowhere near that of Asia's, the Latin middle class has been expanding at a rapid clip, and, compared to emerging nations in Asia, household income is already relatively high in many Latin nations.

I am not saying that the long-distance export of goods and services is a thing of the past—far from it. Such trade will soar in the Next Boom. What I am predicting, however, is that the GT2 era will see this flow become more sophisticated and efficient, and regional trade and manufacturing centers will be a natural evolution. As we saw earlier in this chapter, global trade is expected to triple by 2030. That can only happen if a great deal of innovation is applied at every level of the system, from transportation to finance to engineering and manufacturing.

Meanwhile, as corporations in emerging nations continue to mature and their business interests become more transnational in nature, low-cost production and services will move further down the chain. If governments in lower-cost, less-developed nations such as Egypt, Mexico, Bangladesh and South Africa can rise to the challenge of convincing businesses that they offer rule-of-law and stability, then significant export and service industries will develop in new offshore centers, creating jobs and building a middle class.

Internet Research Tips

Plunkett's Next Boom video for Chapter Three:
www.plunkettresearch.com/NextBoom/Videos

For more details on innovative rural mobile phone services, including an interactive demo, see www.google.co.ug/mobile/sms/#6001

Mobile banking—see mCheck at www.mchek.com, Safaricom's M-PESA at www.safaricom.co.ke/index.php?id=745, and MTN's MobileMoney at www.mtnbanking.co.za

Join in the discussion!

- See the Reading Group Guide in the back of the book.
- Go to Facebook, search for The Next Boom.
- Join The Next Boom group on LinkedIn.

— chapter four —

THE GLOBAL MIDDLE CLASS AND THEIR GROWING MOUNTAIN OF MONEY—FUEL FOR THE BOOM

"He who lives by the crystal ball soon learns to eat ground glass."
-Edgar R. Fielder (American economist)

On the Streets of China: American, Japanese and European Brands

If you have concerns that America's trade with China is a one-way street, a walk through Beijing or Shanghai would change your mind; American brands, companies and icons are everywhere. First of all, these are astonishingly modern cities, teeming with consumers who each have a cellphone in one hand and a Visa card in the other. The airports are vast, modern complexes—state of the art in all concerns. From the Shanghai Pudong Airport, I recently zipped into the city in seven minutes or so, on the new maglev (magnetic levitation) train that runs at 400+ kilometers per hour, built by German companies Siemens and ThyssenKrupp. In Shanghai, the skyline is a panorama of cranes and skyscrapers, with the occasional historic building, such as those on the Bund (Zhongshan Road) area fronting Huangpu River, thrown in for flavor. Freeways and public transportation are as modern as I've seen anywhere else in the world. Ship and barge traffic runs 24/7 on the wide Huangpu. In terms of tons of cargo handled, Shanghai is the world's largest port. (The river itself will soon be the object of a much-needed, multi-billion dollar pollution cleanup project.) Tourists can scoot under the river in a tram car tunnel blazing with psychedelic lighting. On the other side, in Pudong, lies a broad flatland, part of the coast where the mighty Yangtze River reaches the East China Sea. The Chinese have built a stunningly futuristic community in what is called the Pudong New Area, the home of the World Expo 2010. Reminiscent of the Beijing Olympics in terms of cutting-edge architecture, the Expo was designed to attract 60 to 70 million visitors from May through October 2010.

In fact, virtually all of the progress, modernization and development you witness in China today happened in a bit more than 30 years. It was only in December 1978 that new leadership, leaving the dismal later years of Mao's reign behind, began the opening of China to the western world by announcing reforms called the "Four Modernizations." The rebuilding, development and modernization of Beijing and Shanghai happened in only three decades, and these two cities alone are now home to a combined 36 million or so people, nearly equivalent to the entire state of California in terms of population. This is an incredible accomplishment that is still in progress and will be added to for decades to come.

In Beijing, the nation's capital, virtually everything is new outside of the historic Forbidden City palaces. It almost feels like you are in the newest, densest parts of downtown Dallas or Atlanta, but nearly everyone on the street is Chinese. American brands are everywhere. Many of the largest hotels feature American names: Hilton, Sheraton, Radisson, Hyatt and Marriott signs are abundant. Wide, modern streets are jammed with mid-size sedans, minivans and luxury cars, many bearing foreign brands: Mercedes, Audi, Volkswagen, Buick, Cadillac, Toyota. Unlike Shanghai, there are no pollution-spewing motor scooters on the roads of Beijing. The streets are lined with wide sidewalks, and cyclists enjoy broad bicycle lanes on many major thoroughfares. Recently built high rise housing and office buildings abound.

I could wander about for hours in Beijing's modern, multistory shopping malls that seem to be on nearly every corner. In the malls of

Beijing and the shopping districts of Shanghai, I received an extremely entertaining jolt of America's presence in Global Trade 2.0 when I observed the endless stream of stores and restaurants bearing U.S. brands: Polo Ralph Lauren, Tiffany, Starbucks, Nike, Kodak, Sizzler Steakhouse. In addition, I found that many of the local stores are stocked with American brand consumer items: Gillette, Kleenex, Oreos, Ritz crackers, Apple, Coca Cola, Pepsi, Hershey. At the Shanghai airport, I was greeted by a massive color graphic advertising Cadillacs. One of the most dominant buildings on the Huangpu River skyline is occupied by America's Abbott Labs. Among the most interesting visions in central Shanghai is a giant storefront selling NBA-branded items from the U.S. National Basketball Association. China is crazy for basketball and American basketball stars. The malls are also jammed with fashion and jewelry stores featuring European brands: Zegna, Burberry, Hermes, Rolex.

A lot of money is already being earned by American, European and Japanese firms in selling to the Chinese. Multinational manufacturers enjoying high levels of sales include Danone, Coca-Cola and Nestle in the food sector; Kimberly-Clark in paper goods; cosmetics makers L'Oreal, Shiseido and Avon; supermarket and convenience store operator Carrefour; and tech companies HP, Sharp, Samsung and Philips. Goods from these firms intended for local consumption are often manufactured within China, with product modifications and packaging appropriate to local tastes. By 2009, tire maker Goodyear had nearly 800 dealerships in China, operated by local partners, a logical move for a multinational firm that sees great potential in the fact that Chinese buyers will purchase some 19 million new cars in 2011 alone—more than the number of cars that Americans bought during peak years in the mid-2000s boom. Nabisco has 3,000 employees in China, making cookies and crackers with American brands, but modified to meet Chinese tastes.

In fast foods, America's Yum! Brands, owner of KFC, Pizza Hut and Taco Bell, now operates thousands of restaurants throughout China. The Chinese market is so powerful for Yum! that it was opening more than one new store there daily, on average, in 2010. Its KFC unit now has more than 3,000 restaurants in over 650 Chinese cities. By 2010, McDonald's had 1,100 restaurants in China, and business is so promising that the firm ramped up investment there by adding 135 new restaurants for the year, and redecorated most existing units to reflect the new McDonald's café concept. I have recently dropped into McDonald's, KFC, Dairy Queen and Burger King restaurants in locales such as India, Hong Kong, China and Thailand for tasting or observation—and at one point I stumbled into a McDonald's after a hot, sweaty 12-mile hike around Singapore due to the simple fact that I was famished. The food was comparable to that of their stores in western nations, although some local items have been added to menus. Prices were appropriately modest, the service was always a cut above that found in the same restaurants in America, and the level of cleanliness was impressive.

How Consumer Markets Will Grow and Multinational Companies Will Serve Them

Consumer product manufacturers see the ballooning middle class in developing nations as an important part of the ever growing global market. Companies like Unilever and Procter & Gamble are successfully pushing into the frontiers of selling to rural households by creating very low-cost, small packages of soaps, cleansers and personal hygiene items. Meanwhile, in the cities of Shanghai, Mumbai, Beijing and New Delhi, urban household income is growing, and the world's manufacturers and service providers are eagerly seeking market share while they push their brands and establish customer loyalty. By 2014 or so, China will be ahead of Japan as a consumer market, but will remain far behind the U.S., where consumer spending will be about four times the amount in China.[1] Over coming decades, that America-China consumption gap will close as Chinese wages rise dramatically and China becomes home to an ever-larger middle class making ever-larger household purchases. Market growth on this scale simply cannot be ignored by global companies that want to prosper in the future.

This is not to say that foreign firms find it easy to begin selling their goods and services in emerging markets. Obtuse regulations, restrictions on local ownership by foreign firms, tough licensing laws and myriad other roadblocks, not to mention cultural differences, can make entry by foreign businesses into new markets seem next to impossible. Perseverance and careful adoption of both products and business practices to local tastes and regulations, along with careful selection of local partners, will pay off. There is an important, historic Asian precedent. During Japan's days of rapid, post-World War II growth, it was difficult for American and European firms to get a foothold in Japan. Multiple layers of restrictions and regulations often made it difficult to enter the Japanese market. I think back to my first trip to Tokyo in 1980. The Japanese had made serious inroads to the American automobile market by that time, and Hondas and Toyotas were gaining market share, particularly among young buyers. But, back in Japan, while the streets of Tokyo were crowded with relatively new automobiles, I saw only one American car—a single black Cadillac that I suspect belonged to the American embassy or an American corporation (Admittedly, American cars at the time were suffering from poor quality and design. Nonetheless, America's auto manufacturers were complaining honestly about restrictive Japanese import practices.) There were virtually no American brands to be seen on store shelves. Japan's regulations and market conditions made it difficult for an American firm to market goods there on a competitive basis. Similar, daunting difficulties often face foreign firms attempting to establish or buy businesses in India, China and other emerging nations today.

Nonetheless, patient multinational corporations have scored major successes, often in partnership with local firms. For example, Wal-Mart hopes to open huge numbers of stores in India, in partnership with Bharti Enterprises. In May 2009, they opened their first Best Price Modern Wholesale store in Amritsar, similar to a Sam's Club in the U.S. Procter & Gamble is doing $3 billion in business annually in China.[2] Giant drug manufacturer Novartis reported, in 2009, that its sales in China have been growing at a 30% annual rate. It announced plans to invest $1 billion to expand its laboratories in Shanghai. Car maker Audi is selling very successfully in China, where it expected 130,000 unit sales for 2009, while forecasting that China will become its biggest single market, at 250,000 units per year, by 2012 or 2013, outselling its home nation of Germany.[3] Meanwhile, China, where Buicks are particularly popular, became GM's largest market in 2010, and may be its brightest hope for the future.

India's consumers were spending vastly less money per capita than China's, as of 2010, but exceptional growth as a consumer market looms ahead in India nonetheless. In August 2007, the McKinsey Global Institute released a report estimating that, if India continues its recent record of growth, average household incomes will triple over the following two decades, and it will become the world's fifth-largest consumer market by 2025.[4] Here again, firms that want to prosper for the long run cannot ignore this potential.

Despite an Incredible Record of Improvement, Vast Numbers of People Remain in Poverty

A few years ago, labor costs in India and China were exceptional bargains when compared to those in the developed world. However, competition for workers has driven wages higher and higher. Skilled employees in India frequently jump from one employer to the next, lured by higher pay. Chinese manufacturers often complain of shortages of qualified workers.

Of course, the offshoring boom has not touched all residents in India or China equally, despite impressive growth in the middle classes. Income inequality is a significant problem. For example, the Asian Development Bank estimated, in 2006, that 77% of Indians lived on $2 or less daily, a vast number of them earning small livings from agricultural methods that haven't changed much since the British first established trading posts on the Indian subcontinent during the dawning days of the British Empire. India's workforce, estimated at 523 million in 2008 (nearly four times that of America), is woefully underemployed. According to U.S. Central Intelligence Agency estimates, 60% work in agriculture. The suicide rate among farmers has been extremely high, as many fear they are doomed to a life of economic failure. Clearly, there is room for substantial investment in rural education, transportation and employment.

China, since its economic reforms began in 1978, has been a much bigger beneficiary of foreign investment than India has. This is one of many factors that have fostered China's faster growth. China's exports grew ten-fold to nearly $1 trillion annually from 1978 through 2006. Nonetheless, China has hundreds of millions of low-income residents subsisting on low-tech agriculture and remittances from their urban-living children, despite the creation of more than 135 million jobs in that 28-year period. At the same time, personal income has made great strides in China. World Bank figures show that more than 600 million people in China were living on less than $1 per day in 1981. By 2005, that number had dropped to about 150 million, a 75% decline. This is essentially an indication of the number of people who have risen out of abject poverty. By another count, the number of people in China who are living at the more desirable level of $2 to $13 per day soared fourfold from only 174 million in 1990 to 806 million in 2005. These people may not be affluent or able to afford much by Western standards, but they have achieved entry into the lower middle class in local terms. China has the world's largest labor force, at more than 800 million, according to the CIA. Despite the fact that 43% worked in agriculture in 2006, China is a manufacturing powerhouse, with 25% of the workforce engaged in industry. Only 32% worked in service jobs, a very low percentage compared to India, or compared to well-developed nations.

This chapter's discussion has focused on China and India because of their overwhelming numbers in terms of population base, a combined total of about 37% of the people on Earth. However, when considering a global picture of the Next Boom, there are many nations with smaller populations that potentially have very bright futures. Such a list could include Turkey, for example, but it would be topped by Brazil. By some estimates, Brazil will soon grow to be the world's fifth-largest economy overall, surpassing both Britain and France. This could occur as early as 2014-2015. Among the positive trends at work in Brazil are a government that has restrained inflation for more than 10 years (a remarkable achievement in South America, which has a lengthy history of extremely high inflation), exceptional deep water offshore oil fields, an outstanding agriculture sector and abundant fresh water, along with other tremendous mineral resources, a young population and an entrepreneurial spirit. In 2009, the population of Brazil was only about 192 million, but it is growing. By one government forecast, Brazil is expected to peak at a population of 219 million in 2037. Typical of emerging nations, Brazil faces daunting challenges in terms of crime, education, health care and infrastructure, but it is attracting foreign investment in large amounts, while building world-class corporations at an admirable rate. A sterling example is Embraer, the manufacturer of state of the art commuter-size jet aircraft. This highly competitive firm has significant market share in America.

The biggest advances in developing nations are yet to come. While hundreds of millions of people throughout Asia and Latin America will enter the middle class for the first time over the near future, people who are already in the middle class will find their incomes rising substantially. Together, these trends will provide significant fuel for the Next Boom.

Exactly What Is "Middle Class?"

A definition of middle class needs to start with an understanding of "purchasing power parity" or PPP. This concept is used to attempt to account for differences in local prices in a given country such as Mexico, China or Indonesia, compared to prices for similar items in, for example, America. That is, it's an effort to compare apples to apples. PPP is difficult to compute. For instance, to attempt to evaluate an emerging nation's total GDP on this basis, a value, in American dollars, has to be assigned to all goods and services produced in that country. At best, PPP is only an estimate. Analysts use PPP as an educated guess as to how much money someone has to earn, in local currency, to enjoy certain consumer goods or a certain level of lifestyle.

Businesses involved in global trade are paid in real currencies, not in monies adjusted for PPP. Nonetheless, PPP can be a useful tool when attempting to judge the financial status of households in widely disparate nations. Branko Milanovic and Shlomo Yitazki attempted to set a scale for determining the minimum amount of income needed to be middle class in local terms for various nations, on a PPP basis. At the time of their research, Milanovic was at the Research Department of the World Bank, and Yitazki worked at the Hebrew University and the Central Bureau of Statistics, Jerusalem. In their study published in 2002 (*Decomposing World Income Distribution: Does the World Have a Middle Class?*), they determined that the amount of income needed to achieve middle class status would be $12 daily for a resident of Brazil, on the lower end of their scale, and $50 daily for a resident of Italy on the upper end.[5]

The World Bank estimated that residents of emerging nations who were middle class totaled 400 million in 2005.[6] They forecast that this group will grow threefold to 1.2 billion by 2030. According to the World Bank in its report *Global Economic Prospects 2007*, there will be "...average per capita incomes in the developing world of $11,000 in 2030, compared with $4,800 today...Average incomes in rich countries will also increase dramatically: the average income of the children of today's Baby Boomer is likely to be nearly twice that of their parents."[7]

That is, there will be 1.2 billion residents of emerging nations such as India, China and the Philippines who have achieved sufficient income to make at least a small number of discretionary purchases. These people will provide a new market for lower-end Western style goods and will be able to set aside at least a small amount of their income as savings. This is a global step forward that will have immense economic impact. A significant portion of these 1.2 billion people will

GDP by Purchasing Power Parity (PPP), Selected Nations, 2009

(Total GDP in trillions of $ U.S., and GDP per capita. Both numbers are provided on a PPP basis.)

Nation	Total	Per Capita
U.S.	$14.26	$46,400
China	$8.78	$6,600
Japan	$4.13	$32,600
India	$3.56	$3,100
Germany	$2.81	$34,100

Note: Each nation's gross domestic product (GDP), in total (in trillions of U.S. dollars) and per capita (in whole numbers). For the purposes of this table, GDP has been converted to a common currency (the U.S. dollar) and a standard (purchasing power parity—PPP) that attempts to eliminate the differences in price levels between countries.

Source: CIA *World Factbook*, data gathered July 2010

be members of the "upper" middle class, with substantial discretionary funds to spend on consumer products, automobiles and vacations. These newly enlarged consumer markets and relatively affluent middle classes will not happen overnight, but they do have tremendous promise for the 2010 through 2030 period as household income will grow dramatically. Today, the leading consumer markets remain the U.S. and EU by far. However, China was expected to surpass Japan to become the world's second-largest economy in 2010, in terms of real GDP. McKinsey & Company forecasts that China will be the world's second-largest consumer market by 2014, in terms of total consumer spending.[8]

The World Bank estimated that average per capita Chinese incomes would grow from a PPP level equal to only 19% of those enjoyed in high-income nations today, to 42% by 2030. Those are averages—many Chinese consumers will be enjoying upper class incomes equal to those found in wealthy households in America or Europe. (China's booming growth is minting new billionaires at a rapid rate.) That means tens of millions of additional Chinese will be purchasing imported food brands, driving Cadillac and Mercedes automobiles, buying jewelry from Tiffany and Cartier, wearing designer clothes and travelling to resorts in Hawaii and museums in London on vacation. The consumer market potential is gargantuan.

Working for IBM

One offshoring trend that has continued to be strong is the hiring of thousands of engineers and researchers to work in the offshore offices (particularly in India and China) of major tech firms such as IBM and Microsoft. IBM first established a small presence in India in 1992. Soon thereafter, it opened the India Research Lab in 1998 in New Delhi, and later expanded its research efforts to Bangalore. By 2009, IBM's headcount in India had grown to approximately 80,000 in 35 work centers, including people employed in research, consulting and operations. This means that 80,000 Indian citizens enjoy the pay, benefits, training and prestige of working for a world class, U.S.-based technology firm. Such a step up in life could only be dreamed of by most of their parents.

In China, IBM established its China Research Laboratory in September 1995 at the Zhongguancun Software Park in the northwest Beijing area. This lab is focused on the development of technology to aid in the visualization of data. The implications of technology-based job development like this for India and China over the long term are immense, but a nagging question looms in many minds: Are these professional job gains for Asia a major loss for America?

Global Trade, the Displacement of Workers in Developed Nations and How to Boost Competitive Advantage in America

Concerns about offshore displacement of local jobs in the U.S. and other developed nations are valid. It is also true that opportunity flows both ways, and this flow will increase dramatically as Global Trade 2.0 (GT2), and the relatively free flow of commerce among nations, continues to evolve. As the period of gloom and financial turmoil during the Great Recession fades away, this flow will soar above the highest levels of the last boom. This will be a swelling tide of trade in goods, services and intellectual property, plus a constant, multinational flow of investment capital.

Residents of mature, developed economies, such as the U.S. and the U.K., are well aware that developing nations like India and China are booming. Opponents of free trade and globalization believe that the developing world is growing at the expense of everyone else.
Some are terrified that nations in Asia are eager to supply teeming hordes of workers, ready to toil away for low pay in offshore factories,

Population of Selected Nations, July 2010
(In billions)

Nation	Population
China	1.330
India	1.117
U.S.	0.310
Japan	0.126
Germany	0.082

Source: CIA, *The World Factbook*

call centers and software consultancies, thus threatening jobs back home. Some people would prefer to close the gates to imports, while supporting themes of "Buy American" or "Buy Canadian" or "Buy German" goods, depending on where they live. Even the highly successful North American Free Trade Agreement (NAFTA), enabling a steady flow of goods among Canada, Mexico and the U.S., has been a constant target of some political candidates seeking to squeeze an extra vote or two out of a concerned electorate. I recently sat at a dinner table with a retired, "buy American" consumer who studied cars he might purchase in minute detail, attempting to find models with the highest level of domestic U.S. content—not easy in a world where a new car's parts might come from 10 different nations before being assembled in America.

Even the most casual observer of the world of business can readily see that competition for customers, revenues and capital is always intense, and that there are often surprises in who wins and who loses this competition. Firms that falter are flattened by competitors, often foreign competitors. For an entire nation, poor leadership can lead to essentially the same sort of failure. For a nation's economy, being a winner may involve a bit of geographic or geologic luck, but it is also likely to involve foresight, effective strategy, effective government policies, execution and teamwork, with the end goal of enhancing its ability to deliver high-value goods and services to the global marketplace at attractive prices.

In the end, a nation (outside of the failed economies of a handful of countries including Cuba, Myanmar and North Korea) relies on businesses for prosperity, employment and future growth. Significant success in business, and therefore success in job creation, requires at the very least that the following conditions be supplied by the environment in which a company operates, and to a vast extent these conditions are under the direct influence of government:

- Access to capital, including reasonable access to debt
- Access to a workforce with appropriate education and skills
- Tax and regulatory systems that encourage business formation and enterprise
- Effective infrastructure

Global Trade 2.0 is business competition at the highest possible level. "Globalization" and the tools that now enable global trade at blinding speed have turned the playing field upside down. Unfortunately, nations that have been running economies based on high debt and overconsumption have not been well prepared to compete in today's environment of global trade on steroids. This was the position of the United States and many other mature nations as the financial crisis of 2007-2009 ensued. In general, America failed to prepare fully to face a new onslaught of global competitors. As we saw in Chapter Two, Americans overspent, over borrowed and undersaved. Meanwhile, public education failed to live up to the challenges of a more competitive world.

To fare well against global competition, America needs a well-educated workforce. As of October 2008, a study by the Pew Research Center found that slightly less than 40% of Americans in the 18-24 year age bracket were enrolled in either two-year or four-year colleges (up from only 24% in 1973). While this number is a distinct improvement over recent years, it nonetheless indicates a workforce that is not being trained sufficiently to compete in the face of intense global trade. Simply being enrolled in college or technical school isn't enough—graduating with marketable job skills in tune with the realities of the world economy should be the goal. At the very least, a larger number of young Americans need to complete significant vocational training at two-year colleges. In October 2008, a noted expert in education statistics, Mark Schneider, published a paper at the American Enterprise Institute, *The Costs of Failure Factories in American Higher Education.*[9] Schneider, a distinguished professor of political science at the State University of New York (Stony Brook) and the former commissioner of statistics at the U.S. Department of Education, points out that American colleges graduate only about one-half of their students within six years of initial enrollment. He states that

American spending on college-level education, at 2.9% of GDP, is more than twice the average of the OECD nations. Nonetheless, U.S. higher education is failing to churn out a respectable level of graduates—graduates appropriately trained to provide a globally competitive workforce for the future.

The Census Bureau reported that the percentage of 18-24 year old Americans who had dropped out of high school had fallen to 9.3% in 2008 from 19.8% in 1967.[10] This is a welcome improvement, but it still indicates that public education is failing too many of America's 57 million pre-college students.

Asians, to a growing extent, have saved and invested in business and industry, while more and more of their children became engineers, scientists and programmers. They developed, with breathtaking speed, world class corporations like Tata, Evalueserve, LG, WiPro and Huawei that can compete head-on with top corporations headquartered in Europe, Japan and North America. They invested heavily in research and development, good universities and technology parks.

In Order to be Globally Competitive in the Era of a Rising India and China during GT2, the U.S. Should Focus on Curing Current Deficiencies

Positive steps would include:

- Reform of primary and secondary schools, including higher expectations of teachers, students and parents, along with an increased high school graduation rate
- Increased student participation in science, math and engineering
- Increased graduation level of university students
- Greater access to, and utilization of, vocational colleges
- Increased investment in research and development

As we will see in Chapter Nine, education reform (an increased use of education technology plus rapid growth in charter schools) is already trending in a very positive way in the U.S. This is a step in the right direction, with a long, arduous journey left to go. Other daunting challenges remain if America is to regain better competitive advantage. These challenges include the need for a heightened focus on research and development. In particular, an increase is needed in college graduates who are well-schooled in the technical and scientific disciplines that would enable American companies to design, invent and improve more of the technical and intellectual (medical, computer, nanotech, biotech, software, communications) breakthroughs that will create future demand for high-level U.S. goods and services. The development of small businesses needs to be funded and fostered—a few of them will grow to be the next IBMs, Genentechs and Googles. Small business development is critical, as it is the key to high levels of job creation and innovation.

Global trade has taken turns that supporters did not expect, and results have been extremely destructive and disruptive in many cases. The Internet ushered in the biggest surprise of all: Thanks to this low-cost, always-on, 24/7 communications tool, global competition has erupted in services businesses, since the late 1990s in particular. Doctors, engineers, scientists, architects, accountants and other professionals can no longer assume they are safe from low-priced foreign competition—they are not. In *The Next Asia*, Stephen S. Roach, a widely followed expert on Asian business and economics, says it best when he explains the difference that the Internet made to the offshoring of jobs. Before the recent "IT-enabled" phase, when offshoring was primarily in manufactured goods, he says, "For high-wage economies, it was fine to trade away market share in manufactured products. Displaced workers could then seek refuge in nontradable services—incurring steep retraining and other transition costs but eventually drawing security from performing knowledge-intensive tasks..." However, he goes on to explain, "The Internet all but obliterated that sense of job security... With the click of a mouse, the output of a wide range of knowledge workers residing in low-wage developing countries can now be exported to desktops on a real-time basis...."[11] The World Trade Organization (WTO), in its annual report of 2007, estimated that international trade in commercial services was nearly $3 trillion, or about one-fifth of all global trade.[12] To sum up: education, investment, research and development, along with the creation of small businesses, are absolutely vital to America's ability to compete effectively, rather than be flattened in the global market place of the near future. This is as true in service industries as it is in manufacturing.

Demographic Influences on GT2—The Japanese Example

Think back to the biggest lessons of Chapter One: Changes in population and demographics are not evenly spread among the world's nations. Fertility rates are alarmingly low in many countries, far below the replacement rate. Only a handful of nations will account for one-half of the increase in world population by 2050. Aging is dramatically shifting the makeup of the populations of many highly developed countries. Winners include the United States. Losers include Japan, Russia and much of the EU, where populations will be shrinking and aging rapidly over the near future. The implications for GT2 are immense.

Japan has been enjoying status as one of the three largest economies in the world for many years. Household income is high. Unfortunately, the population is aging rapidly, and is forecast to *shrink* 12% by 2050. Although this may eventually change out of dire necessity, immigration historically has not been encouraged, and foreigners or *gaijin* are not readily assimilated into Japanese society. (*Gaijin* may be translated literally as "outside person" or outsider.) The economy has been in slow growth mode for decades, propped up mainly by a government that took on absurd levels of public debt in order to pay for public work projects to keep people employed. The private side of the economy has suffered. Another depressing factor is the nation's near total lack of oil and gas. Oil imports and LNG (liquefied natural gas) imports keep the wheels turning. (Japan was the world's pioneer in the LNG market—a method of transporting natural gas in specialized ships by freezing it to extreme cold temperatures.) In Japan, renewable energy isn't merely a bright idea for the future, it is an economic necessity.

Meanwhile, Japan's aging problem is of stunning proportions. Eventually, by 2050, the dependency ratio, or the number of older people to the number of people in the workforce, will reach the terrifying ratio of one to one. This is a nation in serious trouble. Who will provide needed services? Who will man the manufacturing plants and engineering offices? Who will grow and process food? Who will provide health care? To a growing degree, the answer will lie offshore. GT2 will be boosted by Japan's future needs which, to a significant extent, must be supplied by nations that have younger workforces. This is one of the many reasons that economic ties and trade between Japan and China have already been increasing at breakneck speed, despite territorial tensions over islands claimed by both nations. Workers in China and other nearby Asian nations will be manning factories that make goods for Japanese consumption.

This flow of goods and services to Japan, and the counter flow of the massive retirement savings of Japanese retirees to pay for it, is an example of how GT2 will be based on regional investments and partnerships, the relative age and size of workforces and the costs of shipping. Fifty years ago, communications would have been an impediment. However, the cost, speed and reliability of global communications are no longer issues to worry about. The Internet, satellites, undersea cables and intense competition in the telecommunications industry will increasingly provide always-on, global communication and flow of data at very modest cost. The cost of transportation of goods, on the other hand, may be a significant issue depending on the price of oil, and this may create further bonds in regional trading relationships based on reasonably close proximity.

The Most Important Trends to Watch in Global Trade 2.0

- Growing consumption by the rising middle classes in developing nations
- Continued rapid growth in the services side of international trade
- The influence of demographic and population trends on trade
- Rapid growth in regional centers of trade, regional innovation hubs and regional trade agreements

The Next Billion….

The emerging global middle class means that business and government leaders must start thinking in terms of previously unimaginable numbers—billions in some cases. How can we put some perspective on the looming changes that will result from population growth, rising household affluence, advances in technology and booming global trade?

Let's use the automobile industry as an example. The thought of the next billion automobile owners is either the most intriguing or the most terrifying vision for the near future, depending on how you look at it. In 2010, there were approximately 1 billion cars and trucks on the world's roads and highways. Despite the recent woes of the American automobile industry, future global demand for cars will far outstrip former peaks, creating immense business opportunities. While incomes are rising in the developing world, the price of automobiles is dropping. The most significant development, of course, is Tata Motor's revolutionary Nano automobile, a tiny but effective vehicle that can move four people down the road in relative comfort at a price of only about $2,500. When Tata, one of India's leading corporations, began taking early orders for the car in 2009, over 200,000 people rushed to put down substantial deposits. Now that Tata has lowered the bar on price, largely through innovative engineering and a willingness to create a very Spartan vehicle, the world's other carmakers (Nissan in particular) are rushing to introduce their own low-cost options for buyers in China, India and elsewhere.

In its report *World Economic Outlook April 2008*, the International Monetary Fund shows how, under one method of analysis, the number of cars in emerging and developing economies could increase by 1.9 billion from 2005 to 2050, bringing the world's total to nearly 3 billion automobiles. Will this actually occur? The personal income needed to acquire the cars will be there, but many other questions loom: Will that many consumers find automobile ownership to be desirable? Will public transportation, car sharing systems, commuter trains and other alternatives to individual car ownership reduce demand for personal automobiles? Will fuel, whether gasoline, hydrogen or electricity, be affordable and readily available? Will roads, parking and other traffic infrastructure be adequate to support car ownership on this scale? The same massive inconveniences and costs of individual car ownership that face residents of extremely dense cities such as Tokyo and Manhattan today may dampen desire. On the other hand, new traffic and safety technologies may smooth traffic flow, while highly efficient electric cars topped off by safer nuclear generation plants and concentrated solar plants may turn fuel and pollution into modest problems. In any event, it is reasonable to assume that the world's economies will advance to the point that 3 billion people will desire and be able to pay for access to advanced transportation, whether or not that takes the form of individual automobile ownership on such a massive scale.

Simply put, the trends detailed in this chapter mean that the next billion Visa card holders, Internet users, cellphone subscribers, apartment dwellers, iPod listeners, furniture buyers, mall shoppers, restaurant diners, bank account owners, latte sippers, soda drinkers and international tourists are right around the corner. Growing household incomes, advancing technologies and an evolving global trade platform will generate the economic activity that will produce massive consumer demand in places where, not too many years ago, there were people living in huts with mud floors. This is part of the Next Boom.

Internet Research Tips

Plunkett's Next Boom video for Chapter Four:
www.plunkettresearch.com/NextBoom/Videos

For a fun video of Pudong from the Huangpu River, see "Shanghai World Expo 2010" on YouTube:
www.youtube.com/watch?v=1jn5hdrNx34

The official preview of Shanghai's World Expo 2010 can be seen on YouTube at:
www.youtube.com/watch?v=BX_Wsh3MUpk

Where China's middle class shops: A quick look at a Shanghai shopping mall on YouTube:
www.youtube.com/watch?v=RBOyRcGah4Y

Sikhnet.com: A Visit to India's First Wal-Mart
www.sikhnet.com/news/visit-indias-first-wal-mart-aka-best-price

For in-depth information on PPP (purchasing power parity), see the OECD's web page on this topic: www.oecd.org/std/ppp

Join in the discussion!

- See the Reading Group Guide in the back of the book.
- Go to Facebook, search for The Next Boom.
- Join The Next Boom group on LinkedIn.

— part three —

THINGS

— the next boom —

— chapter five —

COMPETITION, INNOVATION AND ENTREPRENEURSHIP—HOW WE'RE GOING TO SOLVE THE WORLD'S BIGGEST PROBLEMS AND CREATE VAST WEALTH IN THE PROCESS

"It is said that the present is pregnant with the future."

-Voltaire

How Do We Feed Them?

If we're going to add nearly 2.5 billion more people to the world over the long term, and rising incomes are going to push hundreds of millions of them into the middle class, then a few BIG questions regarding these hordes of humans float to the top:

- Can we produce enough food to feed them, and clean water for them to drink?
- Can we create enough energy to power their needs?
- Can we provide them with a reasonable level of health care?
- Can we, without ruining the environment or facing drastic shortages, manufacture the consumer goods and provide the services they will desire?

The answer to these questions is a BIG yes, but a lot of things will change in the process, daunting problems will arise, controversies will ensue, mistakes will be made and a great deal of innovation and entrepreneurship will be needed to make it all work. It could easily require thousands of pages and a lifetime of work to discuss these issues in depth. For the purposes of this book, I am going to spend much of this chapter discussing important aspects of global innovation, entrepreneurship and scientific research, because they are the keys to a prosperous future. To begin with, I will show how this environment of innovation is already laying the groundwork to solve the most basic big question of all: how do we feed them?

Seed Genetics in Paradise

Kauai is a quiet, sparsely populated spot in the chain of islands that make up the American state of Hawaii, one of the remotest places on Earth, nearly 1,900 miles from the closest major land mass. Only 65,000 people live on Kauai, and much of the island is comprised of wilderness preserves, rain forest and the wide, waterfall-enshrouded gorge known as the Waimea Canyon, which is often referred to as the "Grand Canyon of the Pacific." The head of this canyon lies on a mountaintop of volcanic origin in lush Kokee State Park, at an elevation high enough that you can feel a very cool difference in temperature when you drive up from sea level, following a spectacular serpentine road. From Kokee's center, the mountain slopes downward, toward the western shore, until it reaches a sheer drop off at the famous Na Pali Coast, a dramatic, massive seaside cliff that leaves visitors gaping in awe. This landscape represents paradise to some, and business opportunity to others. Not long ago, I hiked along a muddy track from Kokee, down through the forest, to an overlook on the Na Pali cliff tops. Hundreds of feet below me, I could see helicopters and tour boats crisscrossing the waters adjacent to the surf-crowned coast, where tourists on board could enjoy stunning views of the ever-changing colors in the towering cliffs. They would have been surprised if they knew what was going on in the green fields nearby.

Back at sea level, I drove the Kaumualii Highway and then North Nohili Road to the beach near the Na Pali cliffs. This little-used route passes the Barking Sands Pacific Missile Range Facility for miles along the shore, a military base that is highly secure and secretive. If the high fences and military security of the missile tracking station don't scare tourists away, the next set of signs along the road might. At the bitter end of the road lies a deep plain cached between the foot of the mountains and the sea. Here sits a perfect setting for a Michael Crichton novel, a dramatic and otherworldly spot that looks like strange creatures might arise out of the powerful surf, or come swooping off the cliffs onto the green fields below. Notices along the way warn passersby of the criminal penalties they may incur for trespassing, and especially for picking any of the thriving plants that stretch across the carefully tilled fields as far as you can see. Thousands of acres of fertile agricultural soil lie in an area between the radar station, the beach, the foot of the mountain and the Na Pali cliffs, with abundant tropical sunshine, a year-long growing season and vast quantities of water for irrigation streaming from the nearby mountain. You can look, from a distance, but don't touch. Why the intense focus on security? Because this isolated spot is where the future of much of the world's food supply is being created, tested and planned by some of the world's largest corporations. Kauai is ground zero for some of America's most talented geneticists who specialize in agriculture.

"Modified" is Not a Dirty Word

GM is the biotechnology industry's acronym for "genetically modified." The phrase applies to organisms, such as plants, that have been painstakingly, scientifically restructured on a genetic level, in a process often referred to as bio-engineering. Much of the world's most advanced crop science is centered on the use of biotechnology to create new versions of basic food crops: GM seeds for corn, soy, wheat and rice. GM seeds are a relatively young industry, first made commercially available to farmers in the mid-1990s. When you think of Hawaiian agriculture, you may tend to think of pineapples (now grown here mainly for local consumption by tourists, not for export, as much of the pineapple industry has moved to nations with lower labor costs) or perhaps Kona coffee (one of the world's most expensive coffees, and my favorite by far), but the fact is that GM corn for seed is now Hawaii's most interesting crop, and one of the state's major industries. In a late 2008 edition of *Scientific American* magazine, Robynne Boyd reported that Hawaii had already been the site for 2,230 field trials of GM crops.[1] While much of the focus is on corn, these trials have included a search for better wheat, potatoes, rice, cotton and other staples. The world's leading GM plant companies are here: Pioneer Hi-

Bred (part of DuPont), Syngenta and Monsanto. Their trials now cover 4,800 acres, where biologists and their helpers walk their very private fields, sampling, testing and observing. Nearby in their laboratories, geneticists, statisticians and other scientists tabulate field trial results in their effort to create ideal, high-output, hard-to-kill plants.

This is biotechnology in one of its most productive arenas, the modification of the genetic makeup of seeds in order to make plants resistant to insects, capable of fighting off diseases, loaded with nutrients, able to grow with less water and/or much more productive per acre of planting. Crop science has evolved through the decades to the point that, in a good year, a densely populated nation like India can be capable of growing enough grain to feed its hordes of people.[2] Because of rising incomes—leading to more demand for foodstuffs—and a growing global population, a forecast made by analysts at the UN's Food and Agriculture Organization in February 2010 is that agricultural output worldwide needs to increase by 70% by the year 2050.[3] This may place intense demand on the world's supply of water and land for agricultural purposes. On the other hand, biotechnology may well solve much of the problem through the creation of GM seeds that use less water and provide higher output per unit of land. In addition, many other promising technologies and practices will boost agricultural output over the near future. These range from low-cost drip irrigation systems being developed for emerging nations by farsighted entrepreneurs,[4] to better roads for access to food markets (thereby reducing spoilage and waste), to highly efficient professional farm management due to corporate ownership. As we have seen, the adoption of new technologies, such as the use of cellphones on remote farms to access agricultural advice and seek the best buyers for crops, will be a big boost as well.

Much of Africa Will Evolve from a Hunger Spot to a Bread Basket

Investment firms from around the globe are investing in farmland in Africa, and their managers are establishing modern methods and boosting output. Some observers think that Africa has the potential to become a major food supplier to the entire world. Rice, fish from farms ("aquaculture") and crops such as soybeans are becoming lucrative export items in Uganda and elsewhere in East Africa. In Malawi, an innovative government plan for the distribution of fertilizer to farmers has enabled much of that nation's rural population to rise from lives mired in consistent poverty to careers as money-making growers of crops for export. By one estimate, Malawi has come full circle, from importing 40% of its food needs in 2005 to exporting one-half of its home-grown crops in 2009.[5] Lonrho Plc is a multi-faceted corporation that owns businesses ranging from hotels to aviation operations to farms across 17 African nations. Among its businesses is a food processing plant in Johannesburg, South Africa that packs fresh fruit, vegetables and flowers for air shipment to European markets. The crops are grown on the firm's own lands and those of independent farmers in Zambia, Zimbabwe, Malawi and the Democratic Republic of Congo. The Lonrho firm, which decided to reinvent itself from a worldwide collection of businesses to an Africa-focused conglomerate beginning in 2005, may be on the right track. It signed an agreement in June 2008 to develop another major agri-processing center in Malawai and announced plans to rehabilitate 62,000 acres of land in Angola, where it will provide the latest in agricultural management methods. Lonrho's foods currently land on the shelves of such European retailers as Marks & Spencer, Tesco and Carrefour.

Like residents of much of the world, many Africans do not trust GM plants. This will change, slowly but surely. Like the Chinese, the 1 billion inhabitants of Africa have little choice but to adopt higher-output agricultural methods if they are going to feed themselves and take advantage of tremendous opportunities available to their economies from agribusiness.[6] China is setting an example that much of Africa will likely follow. In November 2009, the Chinese Ministry of Agricultural Biosafety Committee issued approval certificates to pest-resistant GM rice.[7] This rice is of the "Bt," or *Bacillus thuringiensis*, variety, which indicates that it contains a naturally occurring, pest-killing bacterium found in soil. (This substance is so safe and effective that organic farmers often spray a solution containing Bt on their crops.) Bioengineers have developed very successful ways to introduce Bt into plant seeds. The bacteria become part of the plant itself, creating an inherent resistance to insects, with tremendous results. This particular strain of Bt rice was created locally at the Huazhong Agricultural University, and is reported to enable an 80% reduction in the use of pesticide while upping crop yield by as much as eight percent. If you stop to consider the economics for a moment, this is huge: an 80% reduction in the cost of one of your most expensive supplies, the pesticide, coupled with a substantial boost in output, times 71 million acres of

Chinese rice farming! Such is the promise of biotechnology when skillfully applied to agriculture. According to the International Rice Research Institute, the Chinese have already more than tripled their rice crop over the past 50 years, largely by improving yield per acre, which is now two-thirds higher than the world's average.[8] The institute estimates that China will need to further boost output by 20% by the year 2030. China recently budgeted $3.5 billion toward GM research on rice, corn and wheat. Positive results from China's massive GM effort may make it more acceptable to other governments to follow suit. China also approved a GM maize in late 2009, which will be an important boost to its ability to raise feed for livestock.

Innovation = Biotechnology Applied to Agriculture = a Better-fed World and Vast Profits for the Food and Agriculture Industries

This is how we will feed the world of the future—through continual enhancement of agricultural technology. Biotechnology is developing the seeds that will enable the farmers of Planet Earth to feed the growing population of the Next Boom. Billions of people will benefit, and billions of dollars in profits will be earned.[9] GM technology is relatively easy to distribute and utilize on a global basis.

ISAAA, the International Service for the Acquisition of Agri-Biotech Applications, published a fascinating study that tracks the global adoption of GM crops from their early days in 1996.[10] By the end of 2009, the group found that worldwide planting of GM seeds had reached 134 million hectares (331 million acres), representing an 80-fold increase over a small beginning 13 years earlier. There was an increase of 9 million hectares in 2009 alone. The largest grower of GM plants by far is the United States at, 64 million hectares, followed by Brazil (21.4 million), Argentina (21.3 million), India (8.4 million, primarily in cotton) and Canada (8.2 million). To look at the numbers in a different manner, consider this: ISAAA reports that 85% of America's maize crop was GM, as were 90% of the cotton crops in America, Australia and South Africa. In total, 14 million farms in 25 nations, ranging from simple family establishments to giant commercial operations, were growing GM crops ranging from soy to corn to cotton. Continued growth in the agribio sector will combine with other farm technologies to enable the world to grow more food and better feed the global population. At the same time, agribio will foster sustainability, including more efficient utilization of water and crop land, and lower use of chemical pesticides.

Crop yields per acre will continue to increase over time as technology becomes more advanced. Due to expiring patents on older products and intense competition from its peers, Monsanto, a leading agribio corporation, is forced to innovate constantly, which is a good thing for the future of food production. The newest corn seeds from Monsanto feature enhanced immunity to herbicides—chemicals used to destroy weeds, but hopefully not the crops themselves. Drought resistance is another focus for upcoming Monsanto seeds.[11] Monsanto recently introduced a soy seed that it hopes will increase yields by seven percent per acre over the previous generation of seeds.[12] Equally exciting, seed engineers are developing plants that will feature significant increases in their nutritional qualities. For example, scientists are engineering a soy plant that contains highly desirable omega-3 fatty acids, thought to greatly reduce the risk of heart disease and provide other health benefits. In other words, you may get the same benefits from eating soy products in the future that you get from salmon today.[13] While GM seed makers set ambitious goals, achieving the desired results can be a difficult matter, and companies in this business have to take a persistent long-term view while continuously supporting their expensive research efforts. In some cases, initial crop yields are not up to the goal. In others, farmers are slow to adopt the high-priced seeds. Also, while many GM seeds have been successfully engineered to resist glyphosate, the herbicide branded by Monsanto as "Roundup," some weeds are developing resistance to this chemical. Nonetheless, within a history of less than two decades, the GM seed industry has achieved tremendous results.

Change = Controversy

The GM plant industry is a good example of the types of challenges that entrepreneurs and scientists will endure as they struggle to help the world move through the trials of the future via technology and innovation. Although scientists have been able to engineer many desirable traits in GM seeds, and the scientific community has largely given GM foods a clean bill of health for years, such modified foods have faced stiff resistance. While many areas of biotechnology are controversial, agricultural biotech has been one of the largest targets for consumer backlash and government intervention in the marketplace. Criticism will continue. For example, a recent documentary film that paints a dark picture of large-scale commercial agriculture, *Food Inc.*, was critical of Monsanto to the extent that the firm published a sharp rebuttal on its web site. Several organic food producers promoted the film, seeing it as a vehicle that will push consumer interest in their non-GM foods. This is exemplary of the wide divergence of opinion about food production—a debate that will not end any time soon. Consumer resistance to food products containing GM material is sometimes fierce. Many consumers in Europe have a strong fear of such foods.

Despite the resistance to GM foods, the need to increase production will not go away, and consumers often respond quickly and violently when food shortages arise. Riots in 2007 and 2008 in a number of Third World countries were fueled by a combination of short supplies and very high prices for staples such as rice and corn meal (this period has been described as the "2007-2008 world food price crisis"). Contributing factors included extremely high transportation fuel costs, booming commodity markets and the unfortunate and unprecedented conversion of a food crop—corn, into a fuel—ethanol, resulting in supply and demand distortions. Price and supply pressures boosted by rising demand for food will eventually make GM foods more acceptable around the world. For example, in Japan and South Korea, a number of manufacturers have begun using sweeteners based on genetically engineered corn for use in soft drinks, snacks and other foods. The manufacturers were looking for alternatives to rising prices for corn starch and corn syrup made from conventionally grown crops. According to Yoon Chang-gyu, director of the Korean Corn Processing Industry Association, non-GM corn cost Korean millers about $450 a metric ton in early 2008, up threefold from $143 in 2006. GM corn was purchased at a cost savings of about 30% per ton.[14]

Syngenta (www.syngenta.com), the result of a merger between the agricultural divisions of two pharmaceutical firms, AstraZeneca and Novartis, is focused on seeds, crop protection products, insecticides and other agricultural needs. Syngenta is in a position to make powerful strategic investments. The firm's annual outlay for research and development is substantial, at about 10% of revenues. Meanwhile, industry leader Monsanto has invested heavily in biotech seed research with terrific results, achieving sales of more than $11.7 billion in 2009. By 2010, the company was facing expiring patents on some of its most important products, Roundup in particular, and its profits are down.

This will force the firm to continue to innovate with newer products, while generic versions of older Monsanto items will reach farmers at lower prices.[15]

A particular concern among farmers in many parts of the world is that GM crops, when they pollinate, may affect neighboring fields, thus triggering unintended modification of DNA in nearby plants. In any event, there is a vast distrust of GM foods in certain locales. America is a noted exception. If you are dining in America, unless you carefully grow all of your own food, you are undoubtedly eating at least a partial GM diet. Meanwhile, a handful of localities in the U.S. have banned or restricted the planting of GM seeds, hoping to protect traditional crops for which local growers are widely known. A typical restriction is to require that GM seeds be planted a certain distance away from non-GM crops.

GM plants and their seeds are highly prized, carefully protected intellectual property, developed at great cost. Consequently, legions of attorneys make sure these seeds are covered by patents. They also draft contracts that the farmers who are end-users must sign, stipulating that the science behind the seeds, and the right to replicate those seeds, is retained by the companies that developed them. Some people have accused Monsanto of being overzealous in pursuing farmers who appear to be using Monsanto-developed seeds without paying for them. The company has also received criticism due to the fact that, at one time, it manufactured toxic chemicals that have risen to varying levels of infamy, such as DDT and Agent Orange. These chemicals have nothing to do with the firm's present business. Unfortunately, anti-GM protestors are sometimes violent or destructive. In the early 2000s, Associated Press reported protests ranging from the bombing of a San Francisco Bay area biotech company, Chiron Corp., to the trashing of a biology lab at Louisiana State University, to extensive destruction of experimental GM plants in France.[16]

Cloned Cows and Nano-enhanced Foods

Meanwhile, food science continues to progress. In early 2008, the U.S. Food and Drug Administration (FDA) declared food derived from cloned cows, pigs and goats to be safe for consumption. In economic terms, this could be a big boost in the near future—the most perfect, healthiest, fastest-growing animals may be consistently reproduced through cloning—the use of genetic technologies to create offspring that are exact duplicates of other animals. The European Food Safety Authority has also declared cloned animal output to be safe, although the EU is talking about new restrictions. A number of food companies, including Smithfield Foods, Inc., Kraft Foods, Inc. and Tyson Foods, Inc., quickly pledged not to use milk or meat from cloned livestock, distancing themselves from what will undoubtedly become a highly controversial matter. Food firms know that a lot of consumer education will be required to make farm animal cloning understood by and acceptable to consumers. In 2010, the possible approval by the FDA of a GM salmon was a hot topic. A firm called AquaBounty Technologies appeared close to winning approval for its farm-raised fish that grow about twice as fast as natural fish. The salmon contain a gene from another fish, and this gene accelerates their growth. This is a technology that may be applied to a wide variety of farm animals in the future.

Nanotechnology is affecting foods as well. As of mid-2010, there were four nano-engineered foods on the market according to The Project on Emerging Nanotechnologies. They were Canola Active Oil, which contains an additive called nanodrops that carry desirable vitamins, minerals and phytochemicals; Maternal Water, which is formulated for gestating mothers and uses colloidal silver ion technology to purify mineral water; Nanotea, which is formulated for an enhanced release of tea essences for taste, as well as an increase in its selenium mineral supplement qualities; and Nanoceuticals Slim Shake Chocolate, a chocolate-flavored diet shake that uses nanoclusters to improve taste and health benefits without the need for added sugar. This is a small beginning to what will eventually be widespread use of nanotechnology in the food industry to do such things as extend shelf life, improve nutritional qualities or create production efficiencies.

Entrepreneurship and Bootstrapped Innovation

The dotcom boom of the late 1990s and early 2000s may have seen wild expenditures by starry-eyed entrepreneurs—huge, Friday afternoon beer busts, Super Bowl ads for firms with no meaningful revenues, expensively decorated office space—but those days are over. Savvy venture capital professionals and angel investors are now much more interested than in the past in efficiency, coaching the founders of startups to be frugal, and fostering cost controls at all levels of the business. In many ways, this is a replay of the 1970s. For confirmation, read Herb Kelleher's great book *Nuts!* , about how he used a modest amount of money combined with revolutionary business techniques to launch what is in many ways one of the most successful transportation companies in the world today, Southwest Airlines. Kelleher's business standards are the techniques of the Next Boom: low costs, disruptive thinking, highly motivated employees, controlled overhead, minimized capital investment, creative use of technology, streamlined workflows and highly effective marketing and branding. In short, this is what I call "needs-based innovation," because entrepreneurs are improvising based on the need to stretch their funds, while at the same time innovating in order to create advanced services, technologies and products. When money is tight and times are tough, entrepreneurs become more innovative than ever.[17] They have to squeeze dollars while maximizing return on investment. This can be a good thing, as these steps often lay the foundation for very sound new businesses.

Entrepreneurs and their startups will be an extremely important part of the Next Boom. I believe that many factors outlined in this book point to a truly exciting era of enhanced entrepreneurship in the near future. I am not alone in that thought. Rich Karlgaard is the respected and entertaining editor of *Forbes*, the business magazine that caters to C-level executives, those who want to be one of those executives, and wealthy investors. He is active on the lecture circuit, and in 2009 his stump speech made comparisons between the entrepreneurs of the 1970s and the innovation he expects to occur today. The '70s were a difficult era, with an oil crisis, high inflation, soaring interest rates, tight credit, high unemployment, the Vietnam War winding down and President Nixon's dirty laundry forcing him from Washington, D.C. U.S. stock values fell 45% from early 1973 to late 1974, Karlgaard reminds us, but it was nonetheless a great time for innovators. He lists the birth of some of today's most outstanding corporations launched during the terrifying mid-1970s, including Federal Express, Southwest Airlines, biotech leader Genentech and technology leaders Microsoft and Apple. Innovation, bootstrapping, creativity and well-thought-out risk taking were the order of the day for entrepreneurs then, as they will be over the near future.

Innovation Will Increasingly Come from Asia

In good times and in bad, some nations are much more accommodating to businesses than others. It can be a bureaucratic nightmare to launch a business in many countries in the EU, due to paperwork, licensing requirements, employment rules, fees and delays. For Americans, there are relatively few restrictions that must be satisfied in order to open most types of service, distribution or manufacturing businesses. However, once a business is established, managers and owners face a rapidly expanding swamp of regulation, taxation and potential litigation that can dampen the innovative spirit and make survival more difficult. It would be reasonable for governments everywhere to observe the job-creating, world-changing success of relatively young companies like Apple, Genentech, Microsoft and Google, and grasp the fact that it is very much in their interest to foster a business environment that is supportive to startups, researchers and investors. Today, many of the world's most exciting young companies are in India and China. Nations that want to flourish and enjoy high levels of employment should vigorously support business development. Otherwise, they will lose their competitive edge in the global marketplace and run the risk of losing out to firms in emerging countries.

Growing investment in, and results from, research and development are exactly what the world needs in order to overcome the challenges of population growth and to foster the Next Boom. Such research is already being promoted very aggressively in a handful of highly competitive nations, and the result will be rapid business growth and high profits. Let's begin with China. For years, some observers have accused China of being incapable of true innovation, relegated instead to a position of imitation and uninspired, low-cost production. This is reminiscent of a line of thought that was incorrectly applied to Japan as it emerged from World War II. Japan evolved into an engineering powerhouse as its industrial base modernized over time, and the same path is likely to be followed by China. In China today, there are three trends at work in the growth of R&D. The first is that the government has an aggressive plan to grow research and development to a level of 2.5% of GDP by 2020, up from about 1.5% in 2010. This would put it about on par with America in terms of R&D as a percent of the national economy. Granted, China has acquired much of its technology base from foreign firms, and still has a long way to go to catch up with the U.S., Japan or Germany in terms of innovation. Nonetheless, the Chinese are poised to become major players in technology development.

Second, many Chinese companies are enjoying escalating success in the global marketplace. Telecommunications equipment makers such as Huawei Technologies, computer hardware makers such as Lenovo, solar cell maker Suntech Power Holdings and a host of growing companies like them are finding increased investments in research to be the key to their continued growth in exports. By mid-2010, Huawei Technologies had 17 research centers around the world, including facilities in Dallas, Texas and Santa Clara, California, as well as a massive $340 million research building opened in May in Shanghai that will eventually house as many as 8,000 engineers. At the same time, Chinese firms are increasing their investments in quality research, design and production in order to create products suitable for the rapidly growing number of middle-class Chinese consumers—China's automobile and personal electronics industries are good examples of this trend. Local markets are an increasingly important focus for China-based companies when they formulate strategies for product development and long term growth.

Third, hundreds of non-Chinese companies of many types have set up serious labs in China. GE alone has more than two dozen research facilities there. Major foreign-owned technology labs include IBM China Research Laboratory, Intel China Research Center, Bell Labs Research China and Motorola China Research Center. This trend is driven by multiple factors, including relatively low operating costs and salaries, the large base of engineers and scientists coming out of Chinese universities and the desire to have labs in close proximity to Chinese manufacturing centers and business markets.

Peking Ducks and Science Parks

In Beijing lies Tsinghua University, a respected institution with the stated strategic goal of attaining rank among the world's best universities. The district surrounding the university paints a useful picture of the contrast between China's old traditions and its modern status in global competition. Tsinghua University dates back to 1911, but upon entering the Tsinghua neighborhood on a recent visit, I drove by tall, recently built buildings bearing the names of top American corporations: Microsoft, Adobe, Sun Microsystems, P&G. I was treated to a feast at a stellar restaurant, part of a business heritage that was founded 150 years ago, *Quan Ju De*, famous for traditional Peking roasted duck. Adjacent to the restaurant, in the Tsinghua Science Park, there was the Google building, home to engineers, researchers and marketers, and the center of deep controversy over whether and how access to the Internet should be regulated. The Chinese government is making a heavy push to move Tsinghua University to the top of the academic hill, and investors, giant corporations and researchers don't want to miss out on resulting opportunities. For example, global business leaders are well aware of the synergies created in Palo Alto, California between a leading U.S. research university, Stanford, and nearby tech company startups and research labs. Billions of dollars of annual business are the result of this synergy, including such offspring as Hewlett-Packard and Google.

There is no let-up in sight for this growth of R&D in China, and it is worthwhile to compare China's research base to that of America. As of 2008, the U.S. had about 1.3 million researchers at work in its labs and science centers. Historically, America could boast of more researchers than any other nation. However, by 2008, China had increased its research efforts to the point that it supported about the same number of researchers as found in the U.S., after doubling its headcount since 1995. If it hasn't already done so by the time you read this paragraph, China's laboratory workforce will soon handily surpass that of America, which will earn China the right to claim the number-one rank among the world's research centers in terms of people employed. This trend is fed by the fact that China is cranking out engineers and Ph.D.-level researchers from its universities at a rapid clip, while those degrees are often unpopular among American students. China produces hundreds of thousands of electrical engineering graduates on the bachelor's level yearly. Also, China is going out of its way to lure home Chinese-born scientists and engineers who have been studying and working in the U.S. and EU. However, there remains a vast difference in total research budgets among nations. While China has hordes of people at work in research, it is far behind more developed nations in terms of the money involved. Even on a PPP basis, America's R&D expenditure was more than two and one-half times as great as China's as of 2010.

Chinese research facilities tend to be located in the same districts as the manufacturing centers that cater to foreign markets: the southern and eastern coastal regions. For example, on the eastern coast near Beijing, you'll find the Zhongguancun Science Park and the Tianjin High-Tech Industrial Park. Further down the coast, near Shanghai, you'll find the Caohejing and Zhangjiang High-Tech Parks. Chinese technology firms are quickly expanding well beyond the fields of consumer electronics and computer and telecom hardware, where China has long had strength. In recent years, the automotive, photovoltaic, health technology, transportation, biotechnology and nanotechnology industries have been developing rapidly there.

There is a Global Focus on Research & Development, and the Competition Will Become Fierce

China certainly isn't the only nation to watch for trends in research and innovation. Growing global demand for technology products and for many types of engineering, coupled with the communications capabilities of the Internet, have launched an R&D boom in many other nations. As in China, some of this research is for locally owned manufacturers, but a great deal of it is conducted as offshoring for companies based elsewhere. Countries with growing research and development bases include Israel, Singapore, India, Taiwan and Korea.

In the South China Sea, bracketed by mainland China, Japan and the Philippines, lies the industrious island nation of Taiwan (the formal name is the Republic of China or ROC—not to be confused with the People's Republic of China or PRC, which is the formal name of China), where total R&D spending was forecast to be over $18 billion in 2010 on a PPP basis. This is a significant amount for a nation with a relatively small population, and the government reports 100 R&D labs operated by domestic companies and another 40 labs operated by foreign firms. Taiwan is on the leading edge of technology-based manufacturing. The country has great expertise, both in the laboratory and on the manufacturing floor, in such sectors as networking gear, semiconductors, computer memory and PC components. Taiwan's researchers are so prolific that they account for more than 5,000 U.S. patent filings yearly. This nation graduates 49,000 scientists and engineers from its universities each year—an amazing number for a country with a total population of a little more than 22 million. Taiwan operates three major science parks containing nearly 800 manufacturers and their labs. This nation has the added advantage of extremely close ties to, and investments in, research and manufacturing centers on the Chinese mainland.

In India, Western drug discovery and manufacturing companies are forging partnerships with local firms at a great rate. For example, Eli Lilly, Amgen and Forest Laboratories have all entered into agreement with Bangalore-based Jubilant Biosys Limited to develop potential candidates for their next blockbuster drugs. This is a big shift from

India's traditional focus on developing low-cost generic versions of Western drugs. Where a few years ago Jubilant Biosys had a team of only 50 in its development labs, it boasted 400 Ph.D.s and Masters-level scientists in 2010. These developments are partly due to the Indian government's decision in 2003 to improve laws protecting the rights of foreign patent holders. At the same time, Western pharmaceutical firms are offering to share intellectual property rights on new drugs as well as a portion of the profits. GlaxoSmithKline was among the first to begin this practice in its partnership with India-based Ranbaxy Laboratories Limited.

South Korea's R&D spending was forecast to be 3.1% of GDP in 2010, and the nation has a goal of increasing that figure to 5.0%. Korean government leaders are focused on increasing research capabilities and basic sciences, particularly at research-oriented universities. In Seoul, the government is backing Digital Media City, a site that it hopes will become a world class hub of developers and entrepreneurs in electronic games, media content and communications technology. The project already houses tenants including LG Telecom, along with broadcasters, creative agencies and startups. The nation hopes that Digital Media City will house 120,000 workers at 2,000 companies as early as 2015.

Nearby in Singapore, the government is backing a technology community known as One North. Its biotechnology area, called Biopolis, is already a large success, with more than 1,000 scientists working in seven research institutes as of 2009. By the end of 2010, new buildings at Biopolis were slated to expand the development to 4.5 million square feet.[18] Eventually, Biopolis may house 5,000 scientists. Another unit of the One North development is called Fusionopolis, a 24-story building housing researchers, designers and entrepreneurs in media, software, communications and entertainment.

The massive growth in Asian R&D will push global competition in technology, physical sciences and life sciences to a new level. This will stimulate the development of new products, technologies and services in vital areas like health care, food production, transportation, energy efficiency and communications.

America's Need to Compete

By nearly any measure, America has long been the world's leader in research and technology. A bit of history may be useful. Organized corporate research efforts began in the chemical dyes industry in Europe in the mid-1800s and soon were launched at a rapid rate across the Atlantic in the U.S. America's can-do spirit, unbridled entrepreneurism and booming economy made fertile ground for researchers. In 1876, at the age of 29, Thomas Alva Edison opened a private laboratory in Menlo Park, New Jersey. Edison's efforts led to his record-setting 1,093 U.S. patents, including those for the phonograph and the incandescent light bulb. Edison's creativity and drive eventually enabled the birth of what would become the General Electric Company, still one of the world's premier research and manufacturing organizations. By 1900, about 40 significant corporate research facilities were operating in the U.S. Rapid acceleration was fostered by a growing middle class that created demand for new products, and was later particularly fueled by the intense demands for leading-edge armaments, transport and other products created for World War II.

During the war, a man named Vannevar Bush was posted as the director of the Office of Scientific Research and Development within the U.S. Government. Bush's mandate was to spur intense innovation that would give America a technological edge in warfare. The results, over an astonishingly short period of time, ranged from advancements in radar to the atomic bomb. Vannevar Bush argued persuasively for a long-term federal commitment to the support of industrial research. By 1962, the federal government was subsidizing nearly 60% of America's corporate research budget, for everything from medical breakthroughs to electronics to defense systems. Thereafter, corporations rapidly increased their own investments in research and development efforts, resulting in a nationwide research base of unprecedented scope and cost. As corporations were striving to compete, the floodgates of research dollars were cranked open, and the American economy was boosted dramatically thanks to true global technology and innovation leadership. Meanwhile, the exciting products rolling out of America's labs, ranging from transistors and semiconductors to better antibiotics and vaccines, made the world a better place for billions of people worldwide.

Depending on the nature of their industries, firms enter into research with widely varying expectations. For example, while aircraft maker Boeing expected to invest about $4 billion in R&D during 2010, this ongoing effort may yield a completely new aircraft model only once every 10 years or so. In contrast, semiconductor maker Intel, which planned to invest about $6.6 billion in research for 2010, introduces breakthrough chip designs on a continual basis. Pharmaceutical makers, accustomed to making immense R&D investments, face a research-to-market cycle of as long as 10 years in order to discover, develop and commercially launch a new drug. In short, some companies, especially in information technology, are driven to expensive R&D efforts in order to compete in their own rapidly evolving industries over the near term, while others patiently invest in long-term R&D hoping it will pay off handsomely through the introduction of entirely new types of products several years hence.

American investment in research and development has been a powerful catalyst. For example, the later part of the 20th Century marked the beginning of the Biotech Era, including the founding of firms like Genentech (launched in 1976 by a venture capitalist and a biochemist) and Amgen (founded in 1980 by a group of scientists). These developments enabled breakthrough drug therapies with high rates of cure. Likewise, there has been a continuing focus on investment in information technology, which generated a leap forward in productivity. More recently, there is growing interest in nanotechnology and a very intense focus on potential advancement in renewable energy and energy conservation. Nonetheless, America's leaders should not take their eyes off the fact that competition outside the U.S. is accelerating dramatically. Growing foreign investments in R&D have the potential to erode America's historic eminence in such technology-driven sectors as aerospace, medical devices, biotechnology, software, oil field services and manufacturing equipment.

On a global basis, spending on R&D has increased rapidly in recent years. In industrialized nations, R&D investment has risen from an average of about 1.5% of gross domestic product in 1980 to about 2.0% today, increasingly close to catching up with the U.S. in these terms. Large numbers of university students around the globe are enrolled in engineering and scientific disciplines—many of them dreaming about potential rewards if their future research efforts become commercialized. Global research collaboration between companies and universities is booming, as is patenting; in fact, it is difficult for patent authorities in the U.S. and elsewhere to keep up with demand. Globalization and cross-national collaboration have such a dramatic effect on research and design that about one-half of all patents granted in America list at least one non-U.S. citizen as a co-inventor.

How to Bet $1.156 Trillion on the Future

The *2010 Global R&D Funding Forecast*, published by Battelle and *R&D Magazine*, estimated global spending on research and development at $1.156 trillion for 2010, up from $1.112 trillion in 2009, on a PPP basis.[19] Their study estimated America's 2010 investment in R&D at $401.9 billion, or 2.85% of GDP, compared to Japan's $142.0 billion (3.41%), China's $141.4 billion (1.50%), and India's $33.3 billion (0.90%). America's lead in terms of total dollars invested, even on a PPP basis, is immense, and Americans should cling to that lead with fierce determination. Looked at another way, the U.S. invests considerably more money in research than the next three countries combined. It also invests much more than Europe's estimated $268.5 billion and 1.69% of GDP. The study estimated U.S. R&D to be 34.8% of world's total.

The bulk of federal support for R&D goes to research in defense and health care. Nonetheless, substantial federal research dollars are expected to flow in 2011 and beyond into such areas as advanced automobile batteries, electronic patient health records and renewable energy. Government research dollars feed projects at universities throughout the U.S. and at many types of private corporations. Major U.S. universities, like the University of Texas and the University of Wisconsin, are eager to patent their inventions and to reap the benefits of commercialized research. Top research universities earn millions of dollars each in yearly royalties on their patents.

Government support for R&D exists at the state level as well. California launched an interesting initiative when voters there approved, in November 2004, $3 billion in stem cell research funding. By 2007, California's stem cell research program was slowly getting underway after combating lawsuits that questioned the authority of the state government to create such a program. Other states across the U.S. quickly began discussing the potential of launching such initiatives of their own. In 2007, voters in the State of Texas approved a $3 billion cancer research initiative spearheaded by former cancer patient and globally recognized athlete Lance Armstrong. The long term result may be heightened competition between tech-savvy states for leading-edge research efforts, at both corporate and university facilities. Meanwhile, U.S. corporations continue to fund massive engineering projects and research budgets of their own. Top research investors among U.S. companies include IBM, Johnson & Johnson, Pfizer, 3M and Apple.

Fundamental Research is Faltering in the U.S.

Despite America's high standing in the world of R&D, warning flags are flying. Adrian Slywotzky, a noted author and expert at consulting firm Oliver Wyman, penned a much talked about article for *Businessweek* in the fall of 2009.[20] He points out an alarming demise of pure, fundamental research and of many of the organizations that funded and produced the most important research breakthroughs of recent times. For example, Bell Labs, founded in 1925 as part of the Bell telephone companies, was once one of the world's stellar research institutions. It has dwindled, partly as a result of the government-mandated breakup of AT&T on antitrust grounds. Eventually, Bell Labs became part of Alcatel-Lucent, which means that this famous, formerly American research organization is now owned by the French. Bell Labs has pulled out of basic physics research, according to a blog from *Wired* magazine in August 2008.[21] All in all, this is a big blow to the U.S. At one time, Bell Labs was an American icon, revered as the place where ground-breaking innovations were launched in such fields as the solar cell, the first transistor (in 1947, leading to a Nobel Prize), lasers, communication satellites, advanced software, the cellular wireless phone system and fiber optics. Slywotzky suggests that new funding of $20 billion yearly could operate three large American research laboratories and five smaller ones. He says, "Split between public and private sources, $20 billion is not so much. If leading companies committed a small percentage of their R&D budgets to pure research in exchange for a tax credit or a government match, a new innovation ecosystem would quickly begin to take shape."

Fully developed nations such as the U.S. have been shifting to knowledge-based economies for decades, as automation took over domestic factory floors (displacing factory workers) and much of manufacturing shifted overseas. In a highly competitive era of information-based businesses, the challenge for developed nations is to maintain their leads in such areas as intellectual property, investment in R&D and university-based research. There is fierce competition among nations to foster advanced education, develop well-trained and motivated workforces, boost productivity and create high incentives for entrepreneurship and investment. Nations that succeed in this regard will invent the new technologies, services, consumer goods and industrial processes that can be sold to businesses and consumers around the world. Put another way, such nations will enjoy the greatest potential growth in employment, income and living standards. Meanwhile, corporations operating in such nations will have the potential to build substantial revenues, earn high profits and create large numbers of desirable jobs.

Reverse Innovation

Innovation that was literally forced by the need to create and operate businesses at low cost and provide services economically is fostering unique new ideas and creating efficiencies around the globe. Bharti Airtel's brilliant creation of a massive cellphone business with extremely low capital expenditures and cheap subscriber fees is a good example. Airtel's business model is being widely studied and copied by cellphone firms around the world. Airtel had no choice in the matter—if it wanted to provide service to India's masses, people who generally had very little money to spend, then Airtel had to create a business that could operate at unheard of efficiency and sell its services at amazingly low prices—needs-based innovation in action. As a result, Airtel created a business based largely on outsourcing, which is easy to imagine since it is based in a world-leading outsourcing nation. Airtel contracted with IBM to operate its entire computer network on an outsourced basis. That is, IBM provides the equipment, the technology and the staff at IBM's expense, and then charges Airtel a reasonable fee for providing ongoing service. Likewise, Airtel contracted with Ericsson and with Nokia Siemens Networks (NSN) to operate its mobile network infrastructure. In many cases, these contracts mean that Airtel was spared the overhead and capital expense associated with initial implementation. Outside companies construct its new infrastructure in each region where Airtel introduces service, and IBM and other outsourcing firms handle customer service. In India, as in many other nations, competing cellphone companies agree to share cellphone antenna towers, rather than duplicating the expense of building and maintaining their own towers. As a result of this focus on innovative cost cutting, highly competitive cellphone firms in India offer good service at rock bottom prices.

In a 2009 volume of the prestigious journal *Harvard Business Review*, Jeffrey Immelt, the CEO of General Electric, wrote that firms like his will see much of their growth begin to benefit from "reverse innovation."[22] This is the concept that needs-based innovation in developing nations, by firms with modest budgets, will lead to breakthroughs in design and revolutionary low-cost products. (The article, *How GE is Disrupting Itself*, was co-written with two professors at Dartmouth's Tuck School of Business, Vijay Govindarajan and Chris Trimble.) Immelt's enthusiasm for reverse innovation is particularly interesting due to the leading role that his firm has long played in research and high-tech manufacturing around the globe. GE is also a powerful example of multinational firms that benefit from global trade. Its revenues generated outside the U.S. grew from $4.8 billion in 1980 to $97 billion in 2008 (53% of revenues). As you will see in a later chapter, GE's own success with reverse innovation includes exciting, lower-cost products from GE Healthcare. Since a primary focus of government and industry in the developed world for the near future will be the restraint of health care costs, the thought of dramatically lowering the cost of medical equipment like this is tantalizing.

The Big Questions Will Receive Big Solutions

Welcome to the near future, where technology will overcome increasingly large obstacles, including the demand for much greater quantities of food, water, energy, transportation and health care. Yes, challenges will abound, and you will hear vast volumes of doom from skeptics. There will be false starts, failures and heartbreaks along the way. Developments in such areas as biotechnology and nanotechnology will lead to significant controversy. Privacy issues, regulatory challenges and ethical questions will loom. Nonetheless, technology, entrepreneurship and innovation have the potential to solve the world's problems, including the challenge of minimizing the environmental impact of an expanding global population, while fostering economic growth and tax revenues for education and support for government debts, along with massive numbers of new jobs for the global workforce. Bill Gates recently stated this well: "Believe me, when somebody's in their entrepreneurial mode—being fanatical, inventing new things—the value they're adding to the world is phenomenal."[23] It remains to be seen which nations will be the most competitive and best realize their potential in this race.

Internet Research Tips

Plunkett's Next Boom video for Chapter Five:
www.plunkettresearch.com/NextBoom/Videos

For a wealth of information on the advance of agricultural biotechnology, see the Knowledge Center at the web site of ISAAA, the International Service for the Acquisition of Agri-Biotech Applications, www.isaaa.org/kc/default.asp

Video: *Global Adoption of Biotech Crops*, from the ISAAA:
www.isaaa.org/resources/videos/globaladoptionofbiotechcrops/default.asp

Biotech fields on the Na Pali coast: To see these agricultural fields from the comfort of your desk, go to Google maps, select the satellite view, and search for "N. Nohili Rd., Kauai, Hawaii."
http://maps.google.com

Wired Magazine's photo essay on the history of innovation at Bell Labs:
www.wired.com/science/discoveries/multimedia/2008/08/gallery_bell_labs?slide=4&slideView=3

Monsanto's rebuttal of the documentary film Food, Inc. www.monsanto.com/foodinc

Join in the discussion!

- See the Reading Group Guide in the back of the book.
- Go to Facebook, search for The Next Boom.
- Join The Next Boom group on LinkedIn.

— chapter six —

ALWAYS ON, ALWAYS WITH YOU AND ALWAYS GETTING SMALLER (OR: BETTER, FASTER, SMALLER, CHEAPER)

"It is far better to foresee, even without certainty, than not to foresee at all."

-Henri Poincare

Buckyballs

Richard Smalley, a Ph.D. chemist, died a hero to science and a hero to Houston's Rice University, where the Smalley Institute continues his work. As the innovations he envisioned come to life over the near future, you may decide that he's one of your heroes, too. Smalley was only 62 when he passed away in October 2005. It is ironic that he died of a cancer, leukemia. I say that because his groundbreaking work in nanotechnology, and the work of people like him such as Rice University's Jim Tour, will lead to better cancer treatments. A story in Houston technology circles goes something like this: In 1996, Smalley was at a scientific conference, where he was scheduled to address a group of physicists. A fellow guest tracks him down and says, "Someone in Europe is trying to reach you." Puzzled, he finally connects with the party by telephone, who informs him he has been awarded the Nobel Prize for Chemistry, thanks to his co-discovery of a new carbon structure, dubbed "C60," a discovery that firmly launched a global rush into nanotechnology research and development. That discovery also put legs under the Next Boom.

Smalley was incredibly bright and focused. Well known in the worlds of physics, chemistry and academia, he traveled constantly, evangelizing about the potential of nanotech. He had a deep belief in the ability of this science to boost the world's economy and to provide solutions to global challenges in areas such as energy and the environment.[1] The Nobel Prize that was shared by Smalley, Harold Kroto of the University of Sussex and Robert Curl, a colleague at Rice University, came from their work that progressed on the Rice campus in 1985. While vaporizing graphite with a laser, they found unusual carbon molecules with 60 atoms each. This was startling, because they were seeing carbon in a crystalline form. The only previously known crystalline carbons were graphite and diamond. Extremely strong and stable (theoretically capable of withstanding a collision with other materials at a speed of up to 20,000 miles per hour), these astonishing new molecules were shaped something like a soccer ball, a multifaceted mixture of 12 pentagons and 20 hexagons. The structure was very similar to the extremely strong and efficient geodesic dome building structures invented by architect Buckminster Fuller in 1967. The scientists named the molecules buckminsterfullerenes, which soon enough became "buckyballs." Their work lit a fire under the scientific and industrial communities that will be transformative over the near future.

Nanotechnology is generally defined as the science of designing, building or utilizing unique structures that are smaller than 100 nanometers in size. We're talking about things that are so small that they are extremely difficult to comprehend. A nanometer is 1 billionth of a meter, or roughly the width of 10 typical atoms. This involves microscopic structures that are no larger than the width of some cell membranes. ("Nano" comes from a Greek word for dwarf or pygmy.) In particular, nanotech often involves the manipulation of materials on the atomic level so that they take on new characteristics, such as increased strength or resistance. On a molecular scale, materials behave differently than they do on a large scale. Novel behavior and novel properties can lead to nano applications that simply aren't possible at large scale. Here's an example that will put nanotech's potential in perspective: A cable made of carbon nanotubes is potentially 100 times the strength of a cable made from steel, but the nanotube-based cable is only one-sixth the weight of the steel cable.

Nanotech is About to Leap Forward like Biotech Did in the 1980s and 1990s

2010 marked the 25th anniversary of the discovery of buckyballs and the launch of the Nanotech Era. Nanotechnology has been in an incubation stage since that beginning in 1985. A vast amount of research has led to deep knowledge on the subject, and to many practical applications—but not as quickly as some people hoped it would happen. The hundreds of companies launched to develop nanotech ideas, some with large amounts of venture capital, have generally been slow to build viable businesses. Investors have often been disappointed. In this regard, the early days of nanotechnology have been very much like the infancy of biotechnology, which eventually grew to become a biotech boom. Growth in the field of biotech eventually led to many uses, including breakthrough drugs and genetically modified seeds, that affect a vast swath of the world's population and have even greater promise for the near future. This slow but sure start of vital new technologies such as biotech and nanotech reminds me of a quote from science writer James Gleick, "I have seen the future, and it is still in the future." However, in the case of nanotechnology's impact on economic growth, the future may not be so far away, as miniaturization in general is already transforming our lives in an ever-present and highly effective manner, and nanotech—miniaturization in its ultimate form—will greatly boost this trend.

Nanotech isn't science fiction—the refrigerator that we purchased for the kitchen in the office at Plunkett Research operates on a cooling system that incorporates nanotechnology. Some of your friends may be wearing pants from L.L. Bean made with nanotech fabric from Oakland, California-based Nano-Tex, a fabric that is virtually impossible to stain. In a few years, our lives are going to be filled with fabrics, paints, drugs, batteries, solar cells, automobile components, building materials, aircraft and computer chips that are enhanced by nanotechnology. A lot of money is going to be made by smart people in this field, and a lot of jobs are going to be created.

The Project on Emerging Nanotechnologies listed more than 1,000 items in its August 2009 inventory of products that have a nanotech component, up from 475 in 2007 and 209 products in their initial

March 2006 inventory.[2] About 60% of the list consists of health and fitness items (such as cosmetics, sunscreens and sporting goods), but the list also includes 80 food and beverage products, such as packaging, 91 home and garden items and 56 electronic and computer products.

One of the more exciting potential aspects of nanotechnology is in the area of self-healing materials. That is, substances such as plastics, paints or metals with nano features that will enable them to automatically repair damage from events such as impact. Nano plastics may be able to change color on a given signal, or change properties on an as-needed basis. Another unique feature of nanotechnology is that scientists are figuring out how to get nano structures to be self-assembling. That is, some molecules can be encouraged to assemble into useful patterns automatically. Researchers at IBM have harnessed self-assembly to create vital, insulating vacuums around nanoscale wires on computer chips, which can enable chips to be produced on an even smaller scale.

Meanwhile, the 2010 Nobel Prize in Physics was awarded to another nanoscale breakthrough in carbon materials. Andre Geim and Konstantin Novoselov of the University of Manchester, England, shared the prize for their experiments involving graphene, which is a remarkable sheet of carbon, only one atom thick. Graphene is strong, essentially transparent and extremely conductive of electricity. It has tremendous potential in a wide variety of technologies such as powerful transistors. Graphene will eventually give the Nanotech Era another powerful boost.

Governments are Backing Nanotech with Billions of Dollars

Governments realize that nanotech is a vital research area. The National Science Foundation reports that the total of funding for such research by all governments worldwide, including America, totaled $825 million in 2000. The next year, the global amount nearly doubled to $1.5 billion. By 2009, in America alone, the U.S. National Nanotechnology Initiative (NNI) received over $1.6 billion in federal funding. In the EU, governments are now committing more than $1 billion in yearly nanotech research funding, and the Japanese government is investing more than $1 billion yearly as well.[3] Analysts at research firm RNCOS forecast that research funding from private firms and government sources combined will top $30 billion in 2013, and that the global market for products incorporating nanotech will reach $1.6 trillion by that year.

MEMS: Tiny Engines and Tiny Switches Enable Big Changes

Another vital miniaturization technology is MEMS or MicroElectroMechanical Systems. You may not know it, but you are protected by a MEMS every time you drive a car that features air bags. A MEMS accelerometer is integral to an air bag, enabling the system to determine when an impact is occurring of sufficient force to require the inflation of the bag. MEMS can be incredibly small and sensitive. They include switches, controllers and even motors and gears that are tiny, roughly 20 to 100 millionths of a meter, but larger than nano scale. They can measure motion, velocity or atmospheric changes in minute amounts. A MEMS accelerometer enables the innovative Wii electronic game machine to react to the body movements of the user.

Miniaturization Is like a Rocket Fuel for Technology and the Next Boom

Since I am making the claim that advancements in technology will power the Next Boom while providing solutions to some of the biggest challenges posed by global population growth, I am going to use this chapter, and the two that follow, to support my case. In these three chapters, I will focus on convergence between leading-edge fields such as nanotechnology, biotechnology and information technology with such realms as energy and health care.

Let's look at a brief timeline for one of the most obvious technology innovations of recent history, the Internet. In the 30 years or so that it has been commercially available, the Internet progressed as follows:

> **First**: "**Sometimes on**." We now look back upon the beginnings of the Internet as a period of exasperatingly slow dialup, during which we used simple modems and telephone lines to connect our PCs with service providers such as America Online.
> **Next: "Always on**." This is when our online experience leapt ahead via broadband subscriptions that continually dropped in price as more and more people signed on, bringing high speed Internet to tens of millions of homes and offices. Broadband, de-

livered largely via DSL or cable modems, vastly faster than dialup, enabled web sites like Amazon.com, YouTube, Facebook and Netflix to offer compelling features. Broadband also enabled sites like online editions of *The Wall Street Journal* and *The New York Times*, as well as MSNBC.com, Hulu.com and ESPN.com to offer instantaneous, 24/7 multimedia access to news and entertainment.

Then: "Always with you." Our online lives became mobile, thanks to 3G cellphones, netbooks, WiFi and myriad other advances that arrived in rapid succession, fueled by miniaturization. Do you want to watch a YouTube video on your iPhone while you wait for the train? No problem, the most popular web sites have versions that are optimized for cellphone screens, and a YouTube application comes standard on the screens of iPhones.

The next progression: "Always around you." This means that nanotechnology is about to come of age, while MEMS and other small technologies will continue to forge ahead. Advancements in miniaturization will enable hardware manufacturers to blanket the world with pervasive, embedded computing. This includes rooms or walls that can sense your presence and turn on lights, entertainment and air conditioning. It also includes smart dust—remote wireless sensors that will analyze everything from the fuel consumption and internal conditions of expensive diesel engines, to the load levels on bridges, to the amount of sunlight hitting a solar energy array.

Over the past two decades, miniaturization has been a vital key to the growth and rapid adoption of fast Internet by hundreds of millions of people around the globe. That is, this growth has been the product of the miniaturization of better, longer-lasting, lighter batteries; the miniaturization of circuitry; the miniaturization of lightweight, bright color screens; and the development of flash memory and incredibly tiny hard drives that can store gigabytes of data and entertainment. Always smaller has been the key to rapid growth in the information technology sector and to soaring economic growth, from the factories of Taiwan to the offices of Apple and Google. This trend will continue. However, miniaturization is expanding far beyond computing, as nanotechnology and other small technologies will enable microscopic engines, more durable materials, and the widespread use of wireless sensors collecting massive amounts of data, 24/7, on everything from traffic, to pollution, to agricultural field conditions, to industrial output.

My 20-Pound Mobile Phone

In 1968, in a state of great excitement and anticipation, I acquired my first mobile telephone. It was made by SCM Melabs of Palo Alto, California. Weighing about 20 pounds and housed in a leather attaché case, it had no dial and suffered from erratic reception. The heavy battery set ran down quickly. In order to make a call, I had to find a radiotelephone frequency that was not currently in use by some other party (we could hear each other's conversations), and then attempt to attract the attention of an operator (by pressing a button and repeating the word "operator" several times) who would dial a call for me. I loved it. I lugged it around everywhere. It cost $2,500 in 1968 dollars. I thought it was a terrific, cutting-edge technology.

Let's jump ahead 22 years in mobile phone history to 1990. The mobile phone I then owned was an immense advancement. It was small enough to fit in my hand, but not in my pocket. Made by Motorola and retailing for about $1,000, it used advanced cellular technology, which meant that I could travel from spot to spot while talking, and my call would be passed along to the nearest "cell" or antenna as needed. I could dial phone calls by using a keypad. Motorola called it a FlipPhone, but one of my neighbors referred to it as "Jack's funny phone," and thought me ostentatious for carrying it around constantly. That was two decades ago.

Although the cellular phone was steadily gaining ground at the time, few people would have believed it if they were told that within a few short years they would be carrying netbook computers, tablet computers and tiny cellphones that could access virtually all of the world's public data and entertainment instantly, in color, no matter where they were, via wireless Internet. The idea that cellphones would replace landlines in many homes, or that e-mail and online payment systems would soon replace envelopes sent via snail mail, would have been hard to believe as well. The new fax machines seemed incredible enough a few decades ago. I had a new Panasonic fax on my desk in the late 1980s, but initially I only knew one other person with a fax. He would send me hand written messages once or twice a week: a drawing of a donut next to a cup of coffee, and a time like 8:00 am, inviting me to coffee—not exactly a productivity enhancement.

Within a few months, however, I found my fax to be in regular use for business purposes as the global network of fax owners expanded at a soaring rate. Meanwhile, long-distance telephone calls had become cheap, thanks to increased competition from upstarts like Sprint, and personal computers were soon widespread. Clearly, trends and technologies were starting to combine in the world of information; prices for devices and services were plummeting while improvements in technology were rolling out in rapid succession. Something big was about to happen.

The Convergence Age: A Mass Market of Information Addicts

The Internet and miniaturization helped to fuel much of the Great Boom of 1982-2007. Thanks to the rapid commercialization and product development that surrounded it, producing immense enhancements to global communications, productivity and collaboration, the Internet also ushered us into what I like to call the "Convergence Age." That is, the convergence of telephony, entertainment (including video and audio) and data of all types onto one digital platform. Initially, this platform was the desktop PC. Today, the platform is increasingly likely to be mobile (a cellphone, a laptop computer, a netbook or an iPad), and in the near future, the platform will also be an Internet-connected TV set in your living room. The very rapid, widespread adoption of broadband brought the Convergence Age into being in the United States once a mass market of 26+ million fast Internet subscribers was reached around the first quarter of 2004.[4] As of early 2010, this market was already at a mature level in America, consisting of approximately 100 million homes and businesses with broadband plus another 50 million wireless broadband subscribers accessing the Internet on their cellphones and other mobile devices. To put it another way, the portable, always on environment that we live in today would not have been possible without the convergence of several technologies more or less at once: the personal computing device, digital telephony, advanced chip technology, lithium-ion batteries, fiber optics and advanced wireless communications, to name but a few.

All of these technologies were dramatically enhanced by the ongoing trend of miniaturization. Without ever-shrinking circuitry, modern network gear would be too slow, too heavy and too expensive; Internet-enabled, pocket-sized cellphones would be impossible; and personal computers would be extremely slow, heavy and uninspiring. My first PC was called a "portable." In reality, it was an anchor. It weighed 25 pounds, had a five-inch black and white monitor, lacked a hard drive, relied on big clumsy floppy disks for media, and cost about $3,000 in 1982 dollars. The only thing that made it "portable" was the fact that the keyboard was hinged so that it could close into something of a cover, and a handle was attached, so that, in theory, I could tote it around handily. Compared to today's computers, it was excruciatingly slow and cumbersome, but it was a terrific boost to my productivity, and I wrote my first book on it, using Lotus 1-2-3 to crunch data and a word processor from a now-defunct firm to create text. Such a clumsy, limited PC dinosaur is hard to imagine as I write these sentences today in front of dual color monitors connected to a modern PC that is connected to a blazingly fast network.

Plunkett's Law of Convergence

Several years ago I made a speech about the most important trends in business and technology to an assembly of researchers at the Austin, Texas headquarters of Hoover's Online. One of my key points was that Americans by the tens of millions would soon progress from dialup Internet access to broadband, and that this would fuel the economy and bring endless new business opportunities. One member of the audience was a nonbeliever. He stood up and challenged me, "Why would so many people be willing to pay for an expensive broadband subscription?" At the time, broadband was still pricey, in the $60 to $100 per month range for home use, but competition and a rapid increase in the number of subscribers would soon drive prices down to $45 (and eventually much lower), at the same time that faster access speeds were being offered, and exciting new services were being launched by web sites of many types. This led me to create the following statement:

> Online consumer usage grows exponentially as broadband access prices decline and more and more Internet devices are adopted—fixed and mobile. This increases demand for new online products and leads to increased offerings of high-value online services and entertainment at reasonable prices.

This virtuous cycle—faster Internet access, lower prices and can't-live-without-them services, continues without interruption today. Nonetheless, convergence is still in its infancy.

Convergence = Better Data = Business Opportunities, Efficiencies and Economic Growth

Convergence, enhanced by further miniaturization, will add fuel to the Next Boom and create massive business opportunities, while helping to create a more efficient, productive and sustainable world for the growing global population. A continued proliferation of information technologies that are fast to launch and inexpensive to acquire will accelerate such convergence—these enabling technologies include cloud computing (the use of remote servers to hold and process data and applications on a low-cost, as-needed basis) and open source software (programs, often provided at no cost, for which the basic code is available to users for manipulation and customization).

Thanks to convergence, we are about to be surrounded by billions of tiny, networked data collectors. This is extremely promising, because the more that businesses, researchers and engineers can rapidly acquire and analyze in-depth data about the world around us, the more efficient and productive the economy can become. Better data leads to new products and services. Better data leads to better marketing. Better data leads to efficient, on-time delivery and availability of goods when and where they are needed. Better data leads to fewer costly mistakes, faster transportation systems, higher levels of energy efficiency and less pollution. Gathering, analyzing and utilizing massive amounts of data is becoming faster and cheaper on a continuous basis.

For example, there will soon be a convergence of GM seeds (biotechnology)—with nanotechnology—with wireless technologies—with enhanced quantitative analysis of massive databases. Robotics will eventually join this trend. The phrase used to describe this convergence is "precision agriculture." PrecisionAg will rely on remote wireless sensors to provide information to farmers on a square meter (or even smaller) basis: Where is irrigation needed or not needed? What is the nutrient level of a given square of soil—is fertilizer needed? Drip systems will deliver needed water and nutrients on demand to exact zones. Farmers of the near future will be able to combine satellite imagery to map and monitor fields, with embedded sensors to determine soil conditions, GPS or laser field markers to provide precision location information to tractors and equipment, and databases to analyze the results of seeds, fertilizers and irrigation. Fertilizer, water and fuel will be used more efficiently. Crops will be planted and harvested on a more exacting schedule. Crop yields per acre will increase while sustainability will be enhanced.

America Leads in Convergence Technologies, but It Can't Afford to Take That Lead for Granted

As with many other recent breakthrough technologies (for example: the transistor, the Internet, GM seeds, the personal computer and biotech drugs), researchers in the United States, like Richard Smalley, took an early lead in the emerging field of nanotechnology and they have held that lead. In a presentation made in 2009, Philip Shapira and Alan Porter, both involved with the Georgia Tech School of Public Policy and the Center for Nanotechnology in Society (Tempe, AZ), showed that America far outpaced China, Japan and Germany in the number of nanotech articles for publication produced from 1990-2006, based on a search of scientific citations.[5] In addition, their search of PATSTAT, a global patent statistics database, found that U.S. nano patenting in 2007 was more than three times that of Japan, Korea, Germany or China. Nonetheless, other nations see the immense potential value of nanotech leadership and are competing very aggressively in this field. The study noted above also showed the interdisciplinary nature of nanotech research, as this science is converging with fields ranging from metals, computers and clinical medicine, to environmental science, cognitive science and geosciences.

Blanketing the Earth with Smart Dust

As 2010 began, convergence, including big advances in nanotechnology, was moving ahead rapidly. In the following discussion, I am going to weave some of the threads together to present a better picture of this progress.

Intel and other firms are working on convergence of MEMS, wireless transmitting/receiving devices, and tiny computer processors (microprocessors embedded with software). In a small but powerful package, such remote sensors can monitor and transmit the stress level or metal fatigue in a highway bridge or an aircraft wing, or monitor manufacturing processes and product quality in a factory. In our age of growing focus on environmental quality, they can be designed to analyze surrounding air for chemicals, pollutants or particles, using lab-on-a-chip technology that already largely exists. Some observers have referred to these wireless sensors as "smart dust," expecting vast quantities of them to be scattered about the Earth as the sensors become smaller and less expensive over the near future. Energy efficiency is going to benefit greatly, particularly in newly built offices and factories. An important use of advanced sensors will be to monitor and control energy efficiency on a room-by-room, or even square meter-by-square meter, basis in large buildings.

In an almost infinite variety of potential efficiency-enhancing applications, artificial intelligence (AI) software could use data gathered from smart dust to forecast needed changes, and robotics or microswitches could then act upon that data, making adjustments in processes automatically. For example, such a system of sensors and controls could make adjustments to the amount of an ingredient being added to the assembly line in a paint factory or food processing plant; increase fresh air flow to a factory room; or adjust air conditioning output in one room while leaving a nearby hallway as is. The ability to monitor conditions such as these 24/7, and provide instant analysis and reporting to engineers, means that potential problems can be deterred, manufacturing defects can be avoided and energy efficiency can be enhanced dramatically. Virtually all industry sectors and processes will benefit.

Thanks to support from the U.S. Department of Transportation, "ITS," Intelligent Transportation Systems, are already beginning to monitor road conditions such as ice, traffic flow and other input to guide truckers and commuters to the fastest routes. Over the near future, new wireless systems, including tiny sensors, will greatly enhance this effort. (This is more important than it might seem. Researchers at Texas A&M University's Texas Transportation Institute estimate that traffic delays cost the U.S. economy $87.2 billion in 2007 alone.[6]) Another transportation sector use will be sensors that will provide constant surveillance of cargo and freight. This means that it will be much more difficult for terrorists to smuggle explosives or nuclear material in a cargo container. It also means that perishable cargo, such as fresh fruit or vegetables, can be constantly monitored for moisture, temperature and ripeness. Intel and other firms have developed methods that enable such remote sensors to bypass the need for internal batteries. Instead, they can run on "power harvesting circuits" that are able to reap power from nearby television signals, FM radio signals, WiFi networks or RFID (radio frequency ID) tag readers.

The Central Nervous System for the Earth (CeNSE)

HP Labs, part of Hewlett-Packard, has proposed an ambitious and far-reaching project called the Central Nervous System for the Earth (CeNSE).[7] Peter Hartwell, a researcher at HP, envisions one trillion nanoscale sensors and actuators embedded in the environment and connected to networks of computers, software and services to exchange and act upon information. While HP acknowledges significant challenges in terms of getting the cost for such sensors down to a miniscule amount, the company sees significant potential benefits. According to HP, Hartwell envisions the use of sensing nodes about the size of a pushpin. These nodes would incorporate the ultimate in wireless sensor technologies, including MEMS accelerometers that can detect even the slightest amount of motion. In vast quantities, they could be stuck to bridges and buildings to warn of structural strains, and embedded along roadsides to monitor traffic, weather and road conditions. Incorporated within everyday electronics, CeNSE nodes might track hospital equipment, sniff out pesticides and pathogens in food, or even "recognize" the person using them and adapt to personal needs. Hartwell is working on a motion and vibration detector sensitive enough to "feel" a heartbeat. The source of that sensitivity is a 5mm-square, three-layer silicon chip that can detect a change in the position of its center of less than 1-billionth the width of a human hair. It is about 1,000 times more sensitive than accelerometers used in a Wii game machine today, according to HP. There is the potential to add sensors for such things as light, temperature, barometric pressure, airflow and humidity. Eventually, such sensing units might be added to mobile phones for broad consumer use. For example, with a wave over food, a sensor might report the presence of salmonella. Continuing advancements in miniaturization over the near future could make HP's bold vision possible.

The Future of Broadband

While the adoption of broadband by Americans has reached major proportions, the FCC estimates that about 4% of U.S. homes are located in rural locations where DSL and cable Internet service are not available. There are additional problems in U.S. broadband distribution. While it is challenging to make an objective comparison of broadband availability from one nation to the next, America consistently ranks poorly in such studies. A study released in October 2009, funded by Cisco and conducted by the Said Business School at the University of Oxford in conjunction with the Universidad de Oviedo, ranked the United States 15th, behind Lithuania, in "Broadband Leadership." The study evaluated such metrics as percent of households with broadband and broadband quality. In March 2010, FCC chairman Julius Genachowski published a proposal, the *National Broadband Plan*, which has implications for the Next Boom. He not only proposes to solve the problems of rural Internet access, he also has a plan called "100 Squared." His thought is to increase Internet services in America so that 100 million homes would enjoy 100 megabit download speed and 50 megabit upload speed by 2020 (compared to an average of about 4 megabits broadband download speed, and relatively slow upload speed, as of 2009).[8] As discussed earlier in this book, America has a very low population density compared to many other nations with developed economies. This has added to the difficulty of encouraging widespread fast Internet access. For example, South Korea, a global leader in access to and use of online services, has a very dense population, most of whom live in urban areas (many in high rise buildings) where the government was able to successfully encourage the commercialization of very high speed access at affordable subscription prices. As a result, consumer use of online communications, entertainment, financial account management and other daily needs are intense—always on is taken for granted. The highest ranking nations in the Broadband Leadership list were South Korea, Japan, Hong Kong, Sweden, the Netherlands and Singapore, in that order.

The *National Broadband Plan's* additional goals include that the U.S. should lead the world in mobile innovation, with the fastest and most extensive wireless networks of any nation, and that every American should be able to use broadband to track and manage their energy usage in real time. The FCC believes that widespread ultrafast broadband can have a transformative effect on American economic development, just as previous advances in communications, such as the telephone, radio and television, stimulated business and the economy. The FCC's report points out myriad potential benefits to education, health care and energy, along with other important segments of the economy. I am not saying that this report's goals would be easy to achieve, or that they are likely to be completely fulfilled anytime soon. However, high speed, extremely reliable Internet access is absolutely vital to future economic growth, and to America's ability to meet the needs of the future in an efficient manner. It would serve America well if the federal government would make leadership in broadband a national priority and encourage private investment in world class Internet access. "Always on" needs to be "always faster."

Growth in broadband subscriptions worldwide is very strong. Analysts at In-Stat estimated that there were 578 million broadband subscribers worldwide by the end of 2009 (both fixed and wireless), and that the number will surpass 1 billion by 2013.[9] Ultra-high-speed Internet connections will revolutionize many types of careers over the near future, both in America and abroad. Telecommuting for workers (such as sales people and customer service reps) is becoming much more common. Professionals such as accountants, attorneys, researchers and engineers are finding it more rewarding to telecommute when they can use broadband to collaborate with coworkers, make rapid downloads and uploads of high definition drawings and videos, and securely store and access critical files at ultra-high-speed and with high resolution. As you will see later in this book, the delivery of education via the Internet is growing very rapidly. Finally, government and public services are much more accessible from the home or office via broadband, without the need to drive to a physical office and stand in line.

The Future: Pervasive Computing and Complete Mobility Will Be Standard

In the U.S., MIT's Project Oxygen is attempting to define the nature of the personal computer of the future, and the way in which people will interact with the computer.[10] It began in the Massachusetts Institute of Technology's Laboratory for Computer Science. The intent of this initiative is to conceptualize new user interfaces that will create natural, constant utilization of information technology. The project states its goal as designing a new system that will be: pervasive—it must be everywhere; embedded—it must live in our world, sensing and affecting it; nomadic—users must be free to move around according to their needs; and always on—it must never shut down or reboot.

The initiative is centered on harnessing speech recognition and video recognition technologies that will have developed to the point that computer data receptors can be embedded in the walls surrounding us, responding to our spoken commands or actions. (This theory will be exemplified in Microsoft's new "Kinect" enhancement to its Xbox 360 game machine. Scheduled for a 2010 launch, Kinect features voice recognition technology, and it responds to the user's hand or body gestures as commands thanks to a sophisticated, built-in camera and advanced software.) As envisioned in Project Oxygen, a portable interface device would provide an ultimate array of personal functions, sort of like a smart phone on steroids. Meanwhile, stationary computers would manage communications with the user in a continuous, seamless fashion. Interfaces would include cameras and microphones that enable the user to communicate with this massive computing power via voice, motion or the handheld unit. The user's routine needs and tasks would be tended to automatically. For example, the user would be recognized when entering a room, and the room's systems would be adjusted to suit the user's preferences as stored in a personal profile. In keeping with the web services trend of Internet-based applications, most of this system's functions would operate by downloading software from the Internet on an as-needed basis. The emphasis on cloud computing that is growing today will be a boost to this vision.

Nanotechnology Will Create Powerful Semiconductors in the Near Future

Silicon, the traditional material for semiconductors, is running out of gas. For years, the process of making faster semiconductors has required a complicated and expensive streamlining of manufacturing techniques for traditional silicon-based chips, with semiconductors becoming faster and faster while manufacturing facilities become more and more expensive. This process has followed the far-seeing prediction made by former Intel CEO Gordon Moore in 1965 that the number of transistors that could fit on a chip (which translates into processing power) would double approximately every 18 months. This prediction came to be called "Moore's Law," and it has held true, more or less, for more than 40 years.

However, our refinement of today's silicon semiconductors faces immense challenges as more transistors are packed on each chip. This is because the circuitry will become so small that it will be unable to conduct electrical currents in an efficient and effective manner using traditional materials. Scientists have already seen this daunting phenomenon in labs, where excess heat and inefficient conductivity via tiny wires create chaos. Nanotechnology, using carbon tubes, promises to solve these problems, making it possible to construct components for computer circuitry atom by atom, creating the ultimate in miniaturization.

In late 2005, global semiconductor industry associations cooperated to publish the International Technology Roadmap for Semiconductors. An updated executive summary of the roadmap was published in December 2009. Nanoscale solutions to future semiconductor restraints are widely discussed in the forecast. While today's smallest transistors may be only a few molecules across, and millions of transistors are clustered on the most powerful chips, industry leaders now envision transistors and nanoswitches as small as a single molecule or perhaps a single electron.[11] Nanotubes and micro-engineered organic materials are the most promising building blocks of the chip industry's future. Researchers at IBM's Almaden Research Center announced in March 2006 that they had successfully constructed a basic electric circuit on a single nanotube.

By 2010, the state of the art in practical chip manufacturing was based on 32 nanometer technology. That is, the ability to build circuitry so small that it is about 1-millionth of an inch wide. Intel, IBM and HP researchers are conducting research that will likely scale chips down to as little as four or five nanometers by 2020. IBM and Intel are working on a new class of transistors named FinFETs, so called due to an area of the transistor that is shaped like a fish fin. The "fin" points vertically away from the plane of silicon in which the transistor is embedded, affording greater density and better insulation.

Nanotechnology Holds the Key to the Ultra-dense Molecular Memory of the Future

"Universal memory" is a catchall phrase used to describe future-generation digital memory storage systems that will be ultra dense and will run with extremely low power usage. Potentially, universal memory could replace today's flash memory, RAM and many other types of memory. In the future, universal memory technology could be based on the use of vast numbers of tiny carbon nanotubes or nanowires, resulting in the storage of trillions of bits of data per square centimeter.

In 2007, Hewlett-Packard announced a startling breakthrough that promises to increase the number of transistors that can be placed on programmable chips by a factor of eight. By using nanowires in a crisscross pattern, creating circuits for storing data or functioning as minute switches, chips can become much more efficient and save on energy consumption. In 2008, a team of researchers at HP Labs announced the development of a switching "memristor," an idea in integrated circuits that theoretically could lay the groundwork over the near future for advanced nano-circuitry, potentially giving a boost to fields that require intense computing power, such as artificial intelligence and synthetic biology. Memristor is slang for memory resistor, an ultra dense concept that opens the door to energy efficient computing systems with memories that continue to hold data after the system powers down. As of 2009, Samsung, Micron Technology and Unity Semiconductor were also working on memristor technologies. In early 2010, HP announced that it had created circuits utilizing memristors as small as 15 nanometers, creating a high level of excitement in the scientific community.

Technology Will Have an Accelerating Impact on the Next Boom, Far Beyond Most People's Expectations

In the millennial year 2000, a frequent statement in scientific circles, used by people trying to quantify the likely technological progress of the 21st Century, went something like this: there will be more advancement in technology in the next 25 years than there was during the past 100, and more advancement over the next 100 years than in all of recorded history. This line of thinking is based largely on the writings of Ray Kurzweil. The author of the insightful 1980s book *The Age of Intelligent Machines*, Kurzweil is a widely followed and controversial scientist who unabashedly reminds his readers of technology's long-term potential. He is a big booster of the potential of artificial intelligence (as a proponent of the idea that supercomputers and eventually even PCs will far surpass the computational power of the human brain, and that the contents and capabilities of the brain may eventually be reverse engineered and digitized). I will grant you that it can be challenging to agree with his thoughts about harnessing our advancing knowledge of biology to achieve extremely long human life (see his *Fantastic Voyage: Live Long Enough to Live Forever,* published in 2004, and

Transcend: Nine Steps to Living Well Forever, published in 2009, both written with Terry Grossman, MD), and I have not adopted his alleged habit of popping as many as 250 nutrient pills daily. Nonetheless, Kurzweil is a keen observer of the power of convergence, which he has referred to as "manifold intertwined technological revolutions," and he provides us with a very provocative analysis of the recent history of technology's growth.

In a 2001 essay, *The Law of Accelerating Returns*,[12] Kurzweil brilliantly points out that we tend to vastly underestimate the technological progress that will occur in the future, because we have linear expectations. That is, we estimate the future based on the present, without factoring in the exponential growth patterns that are clearly followed by technology. (When looking at my first mobile phone in the 1960s, would many people have imagined something as advanced and tiny as the iPhone of a few decades later? I doubt it.) Here is Kurzweil's estimate of technology's progress for the years 2001-2100. "We won't experience 100 years of progress in the 21st century – it will be more like 20,000 years of progress (at today's rate)." Kurzweil created fascinating charts in this essay that plot the exponential growth tracks that were followed by technologies in the 1900s. As I think back to my 20 pound mobile phone of 1968 and my 25 pound "portable" computer of 1982, I am mentally plotting the track of miniaturization from these devices to my iPhone and my laptop computer in 2010. Now, going forward 15, 25, 50 years, I am mentally plotting a track showing the growth and convergence of biotechnology, nanotechnology and information technology. As I follow this line of thought, it is at least possible for me to imagine that Kurzweil's vision of the future rate of technology's progress could be on track.

Internet Research Tips

Plunkett's Next Boom video for Chapter Six:
www.plunkettresearch.com/NextBoom/Videos

Ray Kurzweil's "The Law of Accelerating Returns"
www.kurzweilai.net/articles/art0134.html?printable=1

To learn more about the current state of precision agriculture, see www.precisionag.com

Buckyballs and similar nano structures, see The Buckyball Collection, Florida State University,
http://micro.magnet.fsu.edu/micro/gallery/bucky/bucky.html

MIT's Project Oxygen, with description and videos:
http://oxygen.csail.mit.edu/Overview.html

Online database of products containing nanotechnology components: The Project on Emerging Nanotechnologies,
http://www.nanotechproject.org

Remote sensors: HP Sensing Solutions, YouTube
www.youtube.com/watch?v=UkIasMZOKhI

Artificial Intelligence (AI): If you want to play with a bit of artificial intelligence yourself, visit Ai Research, an artificial intelligence research project. You can download iPhone apps that enable you to create and train your own "Hal" and "MyBot" AI personalities, and you can interact with an AI online chat feature. www.a-i.com

Join in the discussion!

- See the Reading Group Guide in the back of the book.
- Go to Facebook, search for The Next Boom.
- Join The Next Boom group on LinkedIn.

— chapter seven —

ENERGY—WHY THINGS ARE A LOT BETTER THAN YOU MIGHT THINK

"We are called to be architects of the future, not its victims."
-R. Buckminster Fuller

A Newfound Wealth of Natural Gas—George P. Mitchell's Remarkable Gift to Americans

You might assume that the best way to illustrate the effect that advancing technology will have on energy during the Next Boom would start with wind turbines or some other renewable source, but I don't think so. I think it starts with natural gas. Natural gas provides energy directly to tens of millions of American homes and businesses. It is also the fuel source for about 22% of electric power generation in the U.S., and its importance in that sector is growing rapidly. During the late 1990s and early 2000s, it appeared that America was heading for a natural gas disaster. Near-term shortages were forecast, gas prices rocketed upward and plans were laid to import vast amounts of LNG (liquefied natural gas) from the Middle East at immense expense. The U.S. trade deficit would have suffered greatly as America would have become even more dependent on energy imports, and consumers would have endured much higher energy costs. Then, an extraordinary thing happened in the late 2000s: after a brief period of extremely high prices for natural gas (at times, nearly $16 per million BTUs for spot market prices in the fall of 2005,[1] compared to an average of $4.59 during July 2010[2]), America suddenly found itself awash in gas reserves. You might assume that such a breakthrough was the result of the efforts of giant energy firms such as ExxonMobil or Chevron, but that is far from true. To a large extent, this turnaround, an immense and well-timed economic gift to virtually all Americans, was due to the vision and persistence of one man with a strong stomach for risky ventures. The next time you turn on a gas stove or use hot water from a gas-fired heater, you are benefitting from the efforts of entrepreneur, shale gas pioneer and legendary wildcatter George P. Mitchell.

An inspiring business person, Houston-based Mitchell spent decades exploring in North Texas gas fields. He eventually sold his firm to Devon Energy for $3.5 billion in cash and stock in 2001, but not until he was well past 80 years old. Through the years, he applied his vision and drive far beyond the oil fields. His extracurricular real estate activities paint a picture of a man with extraordinary energy and creative passion, and explain a lot about why he was persistent enough to launch the shale gas industry—in his spare time, he led his firm to develop from scratch a model community, The Woodlands, Texas, on a massive ranch he had acquired to the north of Houston that is now home to nearly 90,000 residents. He is reportedly an admirer of iconic architect and visionary Buckminster Fuller.[3]

Mitchell also has deep roots in Galveston, Texas, a once-thriving island community that fell into deep economic decay in the mid-1900s, where he was born to Greek immigrant parents in 1919. This is a classic immigrant-family-makes-good story, worthy of a long book and a generation-spanning movie. By one account, the family name changed to Mitchell when a railroad timekeeper couldn't pronounce the name Paraskivopoulis and decided to call the recently arrived family head "Mike Mitchell."[4] Today, despite a shortage of commerce and frequent battering by hurricanes and tropical storms, Galveston has great wealth in terms of heritage and historic buildings. In its early days, Galveston was the most prominent city and busiest port in Texas. As he became more and more successful in the world of business, George Mitchell acted upon an uncommon desire to help Galveston, some 50 miles south of Houston, recover some of its former glory. Intrigued by Galveston's historic structures myself, I restored a house there in the 1990s, and got it placed on the National Register of Historic Places. My house was built of wood in 1906, stood two very short blocks from the water's edge, and was as strong as a square-rigged ship. Mitchell and his wife had been restoring old properties in Galveston for decades by the time I arrived, and I quickly became enamored of their accomplishments and zest for life. They developed vast new waterfront neighborhoods on the edge of the city, and they owned the biggest hotels.

I was having brunch in the dining room of the historic Hotel Galvez one Sunday, when I saw a gentleman of advancing age wander down the buffet line in slacks and sport shirt. I did a double-take when he sampled an occasional tidbit from the line as he wandered in no particular order or hurry, casually placing items on a small plate he carried about. It was the first time I saw Mitchell, who owned the beautifully restored hotel. Although he was one of the richest men in Texas, he could easily have been mistaken for an ordinary hotel guest. While I was sitting on the boards of various local organizations such as the United Way, I learned how the Mitchells quietly wrote large checks for nonprofit projects throughout Galveston, often taking no credit. I also learned how it had a been a long and difficult road to his incredible success in the production of unconventional gas, that is, gas from fields where the costs, difficulties and risks of failure were very high.

The Man Who Wouldn't Give Up

George Mitchell was largely responsible for the rapid growth of a massive, independent oil and gas enterprise, Mitchell Energy and Development. The Mitchell family was able to maintain a controlling interest in the firm as a combination of good luck, good timing, steel nerves and superb decisions in the oil field moved them ahead. He is generally credited with developing the techniques, as well as having the courage and tenacity, needed to get gas from the Barnett Shale, the first big "shale gas" production area. The Barnett surrounds Fort Worth, Texas and includes 20 nearby counties. It has made many a North Texas landholder very wealthy. Covering 5,000 square miles, the obstinate Barnett pay zone lies about one and one-half miles below the surface of the Texas plains. In 1981, Mitchell drilled his first well in the Barnett, the C.W. Slay #1, in Wise County, Texas,[5] halfway between Forth Worth and hot, dry Wichita Falls. Many people were convinced that Mitchell would never make a dime out of the tight shale formation, and gas prices, vastly lower than today, did not encourage high-risk drilling. Nonetheless, George Mitchell and his crew were tantalized by geology reports that clearly showed massive amounts of gas trapped in the shale. The problem was: how to get that gas to emerge and flow to the surface at reasonable cost?

Mitchell and his engineers persevered in their experiments in the shale for some 17 years, finally determining the right methods by 1998, to the point that the shale rock formations began to produce gas in good volume. By this time, Mitchell was in his late 70s, but he moved ahead with one test well after another until he was able to demonstrate that the use of certain technologies could turn a dull, money-losing well in shale into a gusher of gas. With relative success then in hand, the Mitchell firm continued its quest to make the entire process more efficient and profitable. It would be difficult to overstate the importance of Mitchell's accomplishment. In June 2010, the Gas Technology Institute awarded him a special lifetime achievement award at a major conference in Amsterdam. By mid-2008, there were 10,564 wells in the Barnett Shale. According to the American Petroleum Institute, the Barnett alone was producing about 6% of all natural gas gathered in the lower 48 states as of early 2010.[6] Many of these wells are actually within the city limits of Forth Worth, under homes, schools, churches and highways, quietly making money for all concerned. Engineers on the rig use computers and advanced software, first to send the drill on its long, downward path, and then to turn the drill at the correct vertical depth and continue it on horizontally with great accuracy. However, it took much more than horizontal drilling to make the Barnett Shale truly viable. First, Mitchell had to uncover the secret—the extensive processes that had to be performed after the holes were drilled in order to coax the gas to the surface.

Although it has been produced in substantial quantity for only a few years, shale gas (now produced in numerous important fields across America in addition to the Barnett) already accounted for about 20% of America's natural gas supply by 2010. By some estimates it could account for 50% by 2035.[7] There are multiple reasons why this is important: First, this shale gas is a domestic supply, not at all reliant on imports. Next, natural gas is easy to distribute. A massive pipeline and urban supply system is already in place. Finally, natural gas has about one-half the carbon impact of coal, and it is relatively easy and inexpensive for electric power plants to switch from relying on coal as a fuel to using gas. Numerous shale gas areas are widespread across America. For example, there is the massive Marcellus shale field and the adjacent Utica and Devonian, running from New York southwest through several states to Kentucky and Tennessee. The New Albany is being actively explored in Illinois and Indiana; the Fayetteville is in Arkansas; and the Bakken and Gammon are in the Dakotas and Montana, to name just a few of the known fields. Tax revenues from shale gas production will soon help to bail out the budgets of many state governments. Depending on whom you ask, shale discoveries have boosted U.S. reserves to between 2,000 and 3,000 trillion cubic feet (TCF),[8] much more than a 100-year supply. (Meanwhile, even more gas is on the way. New pipelines are planned to bring conventional Alaskan gas to the lower 48 states, while massive fields of gas await a resumption of exploration in the prolific Gulf of Mexico.)

Shale gas is not limited to the United States; potentially large formations are already being investigated in areas including Europe and China. A new gas supply of this magnitude is absolutely disruptive—it will bring immense changes to the entire energy sector. Oil man T. Boone Pickens sees gas as an alternative fuel for vehicles. Based in Fort Worth, Texas, Pickens is an outspoken octogenarian of great wealth. He sees the potential of America's abundant natural gas as a transportation fuel to be so great that, by one report, he has invested $62 million of his fortune, along with a lot of his time, in spreading his vision of how natural gas-powered vehicles might dramatically reduce America's dependence on imported oil.[9]

Convergence in the Oil and Gas Fields

Why were we able to tap these immense shale gas resources in a relatively short period of time once Mitchell had shown the way forward? Technology and convergence were the enablers. Software enhancements have combined with much better seismic acquisition to create extremely accurate data about geological structures. Systems like Schlumberger's SimMAP Diagnostics software and its PeriScope 15 directional, deep imaging system give geophysicists an exceptional ability to analyze potential pay zones and guide drillers with pinpoint accuracy. These recent advances in technology (including the miniaturization of electronics converging with advancements in software)

are critically important because a drill path error could ruin a drilling project. The process is complicated, but modern technology makes it possible.

Drillers utilize directional drilling specialists whose computerized controls guide the drilling effort. A drilling site may serve only one horizontal bore that comes off a primary vertical hole, or multiple horizontal holes spreading out over long distances in many directions. This is vastly different from older technology where one rig drilled one vertical hole. Today, once the horizontal holes are completed in a modern shale project, a process known as "fracing" (hydraulic fracturing of the shale formation when water, sand and chemicals are pumped in under extremely high pressure) opens up the structure so that billions of small encasements of gas can escape. George Mitchell and his engineers advanced the use of such post-drilling procedures in the shale gas field. Today, complex computer models and massive databases of geologic data are used to design the fracing for best results. Consequently, from 2000 to 2010, America rose from being a gas beggar to sitting on a wealth of natural gas that plummeted in price while supply soared. Following Mitchell's Barnett discoveries, many once-small shale gas pioneers such as Chesapeake Energy and XTO Energy were encouraged to take the risks that created a new flood of gas to American markets. By betting its future on exploration for shale gas, Oklahoma-based Chesapeake Energy grew from its 10 person beginnings in 1989 to join America's top ranks of natural gas producers, with thousands of employees.

Like all other energy production or generation, shale gas has created environmental controversy. Everyone needs energy; no one wants to live near its source. I am not insensitive to this issue. When I took a break from writing this paragraph, I strolled out to the back yard of my country property where I could see a recently installed rig that was drilling through Texas shale. Unfortunately, if they hit gas, I won't participate in the stream of income. I am simply a neighbor who gets to live with this new activity while I continue to enjoy my modest natural gas bills, held down by low market prices. Nearby, I can stand on a hilltop and stare at the multiple smoke stacks of a coal-fired electric generation plant a few miles away. Neighbors still talk about how upset people were when they learned the plant would be constructed there, decades ago.

In particular, the fact that drilling will penetrate local water tables, and that various steps in the process utilize chemicals that might infest water supplies, prompts deep concern about water contamination. Furthermore, fracing a well can require an input of 3 to 4 million gallons of water. A large portion of the water is recovered, which creates the additional issue of how to treat and dispose of it. Drilling and fracing near residences, farms and water supplies is creating a storm of protest. These challenges are ripe for America's scientists and entrepreneurs to solve. Not long ago, I had dinner with a Ph.D. chemist who is deeply involved in research to improve the safety and efficiency of the chemicals used in such wells. Downhole procedures and technologies exist that, when properly applied, protect water tables. Meanwhile, big advancements are being made in the design of environmentally friendly chemicals for use in fracing. One major issue is which chemicals to use in order to keep harmful bacteria from tainting a well. Major oil field services firms such as Schlumberger and Halliburton are developing greener methods of controlling bacteria, such as the use of ultraviolet light at the surface, or non-toxic chemicals used downhole.[10] This research isn't limited to major corporations. For example, entrepreneurs at a startup, Ecosphere Technologies, have also developed innovative solutions, backed partly by funding from former NFL stars. Meanwhile, technologies for better water salvage and reuse at the drilling site are coming forward.

The Houston Chronicle once described Mitchell as "an ecoconscious oilman before it was cool."[11] He has actively encouraged energy conservation, and he established a series of conferences at The Woodlands to address sustainability. The Mitchell family underwrote a widely read 1999 report from the National Academies, "Our Common Journey: A Transition Toward Sustainability," and they made a $10 million gift to the National Academies to endow future studies in sustainability science.[12] The Mitchells have long been believers in the ability of technology to address major challenges, as evidenced by their $35 million donation to Texas A&M University to greatly enhance its physics department.

Conservation—How Your Refrigerator is Helping to Make a Better World

The broad field of energy over the near future will be focused as much on conservation and efficiency as on the development of new energy sources such as shale gas. Conservation is where the low-hanging fruit lies, and future answers to the energy challenge will be found in areas as diverse as highly efficient automobiles that virtually drive themselves, lighter aircraft bodies and changes in building materials. Convergence and miniaturization will guide these efforts.

The electric utilities industry has told us for decades that it is a lot easier and cheaper to conserve electricity through the use of efficient industrial systems, buildings and appliances than it is to build more capacity to generate additional power. However, conservation is not an immediate fix; instead, it is a long-term evolution. For example, a few decades ago, one of the major expenders of energy in a typical American home was the gas pilot light, burning 24/7 on furnaces, cooking stoves and water heaters. Today's appliances don't have pilot lights; they have on-demand electric igniters, so that no gas is burned at all when the appliance is idle. Likewise, today's refrigerators use about 75% less electricity than the refrigerators of 1975, while holding 20% more capacity, because they feature better insulation and more efficient cooling systems. These are good examples of simple, extremely cost-effective reductions in energy usage, but such changes take time—we didn't see old-technology refrigerators tossed out of all homes in America at once.

Moving up the chain, the concept of co-generation (or CHP, "combined heat and power") is a simple, relatively low tech method to capture and reuse factory heat that is the result of industrial processes. That salvaged heat may be used in any of several ways to power a turbine that creates electricity. The electricity can then be used by the factory, sold to the grid, or both. Ever since the dawn of the Industrial Revolution, factories have been burning such fuels as coal and natural gas to make steam, flame their furnaces and turn their engines, but historically they let the resulting excess heat escape through stacks. This is now less and less likely to be the case. Likewise, for decades oil fields flared off excess gas in brilliant, multi-story towers of flame, even in Alaska, relatively close to the lower 48 states' gas-hungry consumers. Today, except in the remotest fields, that is less likely to happen, as investments have been made in gathering systems and pipelines to bring the gas to market.

Energy Intensity—How Efficiency Rises Along with Economic Development

Energy efficiency advances at the creeping pace of a turtle, but over the years the compounding results are exceptional. "Energy intensity" refers to the amount of energy required for a nation to produce a unit of GDP. The lower the energy intensity score, the better. A nation's goal should be to make intensity as low as possible. In constant dollars (adjusted for inflation and expressed as year 2005 dollars), the U.S. economy has grown from $1.84 trillion in GDP in 1949 to $13.3 trillion during 2008—an increase of 623%. During the same period, America's annual energy consumption rose from 31.98 quadrillion BTU to 99.4 quadrillion BTU—an increase of only 210%.[13] Energy consumption per dollar of GDP (on the same constant, year 2005 dollar basis) dropped from 17.34 thousand BTU to 7.47 thousand BTU.

In other words, after removing any distortion caused by inflation, America required only 43.5% as much energy to create a dollar of economic output in 2008 as it required shortly after the close of World War II, while America's energy intensity improved by a factor of 2.29 times. A table of this progress, as published by the U.S. Energy Information Administration (EIA), shows steady improvements in energy intensity, year by year, for the past 60 years. This fall in energy used per unit of economic output is not limited to America by any means, but is more of a global phenomenon. According to EIA data, China, by some measures the world's largest consumer of energy as of 2009,[14] and long a major concern in terms of pollution and emissions, is showing steady improvement, cutting its energy intensity by about 50% from 1980 through 2004.

In recent years, annual growth in energy usage has slowed dramatically in developed nations, while efficiency has soared. The challenge is to make efficient technologies inexpensive, widespread and readily adoptable in emerging nations. I'm not claiming that this will be easy, or that there will not be big increases in the total demand for energy as the world's middle classes grow and emerging nations become more industrial—far from it. The EIA projects global demand for energy to balloon from 508 quadrillion BTU in 2008 to 678 quadrillion BTU in 2030. This is a 33% increase, most of which will occur in emerging nations, particularly India and China. Nonetheless, I believe that accelerating improvements in energy technologies have a sound chance of reducing that demand, lessening the impact of emissions and greatly boosting efficiency.

Energy: And Now for the Good News

The EIA's *Annual Energy Outlook 2010* includes forecasts to the year

2035. The organization paints a very encouraging picture in this report.[15]

Despite America's projected population boom from 304 million in 2008 to 389 million in 2035 (a 27.9% increase):

- Growth in electricity usage will average only 1% per year during the 2008-2035 period, thanks to the effects of higher prices and improved efficiency, among other factors (compared to annual growth of 7.3% in the 1960s and 2.4% in the 1990s).
- Renewables will climb to a 17% market share of electric generation (from 9.1% in 2008), while coal will decline in share.
- Non-hydropower renewable sources will meet 41% of total electricity generation growth.
- Growth in energy-related CO2 will be only 0.3% yearly, and this estimate assumes there are no new regulatory policies in place.
- Energy intensity will continue to improve substantially.
- Oil usage will remain essentially flat, while reliance on imports of liquid fuels (primarily petroleum) will decrease to 45% of total supply from 57%.

Why do the numbers look so good? Because of the effects of innovation, improved technology and enhanced conservation. In fact, I believe that there will be such an intense, entrepreneurial focus on energy supply and efficiency that the actual results will be much better than these estimates. Thanks to the convergence of multiple technologies, the near future will see tremendous breakthroughs in efficient transportation, enhanced oil and gas production, cleaner electric generation, renewables and power conservation. Nanotechnology will play a powerful role in this trend. Michael Moe, a prolific writer and highly regarded analyst of technology stocks, said in a March 2010 edition of his *Next Up!* newsletter, "Nanotechnology…has the opportunity to disrupt the energy industry, and almost every other industry, as it changes the rules for physics, biology and chemistry."

Convergence will solve many of the problems facing firms that want to produce more oil and gas from difficult formations. For example, in the first commercial application of its CeNSE technology, discussed in the previous chapter, HP and oil industry giant Shell will build a next-generation wireless sensing system to acquire high-resolution seismic data. Data and imaging are everything to the oil and gas exploration business. Better sensors like those based on CeNSE will boost accuracy. This means a much better chance for Shell to drill the right hole in the right place with maximum efficiency and maximum potential production. Nanotechnology was slowly entering the oil field as 2010 began. Nanocomposites, Inc., in The Woodlands, Texas, uses nanotechnology licensed from Rice University to enhance the performance of elastometers (materials with high flexibility or rubber-like characteristics). Its products have already proven to be effective for use in offshore well blowout preventers and other oil field uses. Advanced remote sensors and nanotechnology-based components in blowout preventers may help to avoid future failures like that on the infamous Deepwater Horizon in 2010. An exciting variety of new applications for nanotechnology in the oil field will emerge soon.

Within the energy arena, boosts from nanotech and convergence will not be limited to oil and gas exploration. In fact, the list of potential nanotech uses to enhance energy production, storage and conservation is very long indeed. For example, a lot of time and money is being invested in research using nanotube technology to create highly efficient energy storage devices—essentially giant batteries. Success could bring a significant breakthrough for the solar and wind energy industries, where storage solutions are vital to making alternative power generation more viable. Cost-effective ways to store electricity would mean that wind power could be captured when the wind is blowing and utilized later, and solar power could likewise be banked. A global oil industry conference on nanotechnology was held in November 2009.[16] Discussion at the conference included the application of nanotech to such areas as improved drilling (for instance, the ability to withstand harsh environments, high temperatures and the high pressures of deep wells), drilling fluids, "smart" drill bits, enhanced measurement and monitoring methods downhole, long-lasting coatings and improved post-drilling water filtration.

Running Out of Oil Again? Probably Not

The human capacity for pessimism is amazingly powerful and persistent. Fortunately, when millions of people are angst-ridden about a particular problem, at least one or two entrepreneurs are likely to be a few steps ahead of them—working on innovative solutions. Entrepreneurs know instinctively that increased demand and potential shortages create opportunities, and the field of energy is no exception to this rule. In fact, the energy industry has a long history of entrepreneurs facing high risks, sometimes reaping vast rewards in return for their efforts.

Ever since William Hart dug America's first successful gas well in Fredonia, New York in 1821,[17] and "Colonel" Edwin Drake drilled the first true U.S. oil well in the state of Pennsylvania in 1859, the ability of oil and natural gas to power electric generation plants, transportation, homes and industry has created both immense economic advances and significant controversy. Many times it has been assumed that the world would quickly run out of oil. In 1939, the U.S. Department of the Interior warned that America's oil reserves totaled only enough to fuel the nation for about 13 years.[18] Similar misjudgments were announced on a regular basis in the mid to late 1900s by the federal government and by a continuing stream of respected reports and books by various authors. In fact, rather than becoming scarcer over time, energy, including oil and gas, became much more plentiful. Energy prices can fluctuate wildly. Nonetheless, over much of history the trend has often been lower prices on an inflation-adjusted basis, when a combination of advancing technologies, determined entrepreneurs and alternative sources exponentially expanded the total amount of energy and reserves available for consumption. The breakthrough in shale gas in recent years is a perfect example.

An estimate of crude oil resources on a global basis, published by Cambridge Energy Research Associates in 2006, was 4.82 trillion barrels—enough to take care of the world's needs for more than 100 years. This number included oil shale and other sources that are relatively difficult to tap. Technologies will continue to be enhanced, enabling the recovery of significant portions of these resources, as long as the market price of energy is high enough to justify necessary investments in technology, exploration, development, production and distribution.

High prices for oil in the mid-2000s put a new emphasis on production from alternative, "unconventional," oil sources such as tar sands in Canada (a geologic phenomenon that holds vast stores of tar-like heavy oil that is mined and then treated at great cost to make it usable in a commercial manner) and, in the U.S., oil shale (underground layers that hold a huge treasure of hydrocarbons in a form called kerogen that can yield oil, but are extremely difficult to develop for commercial production). Such fields are significantly more expensive to produce than conventional structures, but they hold immense reserves. Canada's tar sands are so massive that, in terms of total reserves, they are second only to the vast quantities of oil that lie under the deserts of Saudi Arabia.

Meanwhile offshore, highly sophisticated rigs are drilling ever deeper to tap massive reservoirs, using technologies that enable the rigs to go to depths undreamed of 30 years ago. Vast new investments in very deep wells offshore of Africa, South America and elsewhere will bring significant new production to the market over the near future.

Brazil's Santos Basin—Drilling Deep

Mammoth new oil and gas fields hold promise in many parts of the world. For example, from the fall of 2007 through the spring of 2008, Petrobras, Brazil's government-controlled oil company, announced the discovery of three mega-giant oilfields offshore, including a major gas deposit named Jupiter, off the coast of the state of Rio de Janeiro, and an immense new oil discovery about 155 miles offshore of the state of Sao Paulo in the Santos Basin. This is close to the major offshore find known as Tupi, which was discovered in 2006 in the Santos Basin. Combined, these finds turned Brazil into an offshore oil and gas giant almost overnight. Tupi ranked as the world's largest find since 2000 and the biggest in the Western Hemisphere in 30 years. The Santos Basin is a vast area in very deep water. There, oil reservoirs lie under a thick layer of salt, which creates technical challenges due to heat, pressure and potential movement of the salt structure. Tupi may hold between 5 billion and 8 billion barrels of oil, and the latest Santos Basin discovery may hold as much as 33 billion barrels. Drilling a single well in the Santos Basin requires an investment of about $250 million. These wells have an average depth of about 22,000 feet, including 7,000 feet of water. Under the water is the 6,500-foot-thick, corrosive layer of salt. When a drill bit finally penetrates the salt, it can access a pool of oil estimated to run from 250 to 400 feet thick. The total Santos Basin covers about 1,000 square miles. The exploration and production challenges here are similar to those faced on a regular basis in the Gulf of Mexico. Oil and gas fields in Brazil are so promising that Petrobras has set a stunning $224 billion investment budget for the 2010-2014 period.

On "Mars," Bouncing Back from a Hurricane

Meanwhile, back in the Gulf of Mexico, the name Mars refers to more than a heavenly body. On Earth, it is the name given to a floating platform off the coast of Louisiana which produces about 190,000 barrels of oil and 200 million cubic feet of natural gas per day. The oil alone would be worth $14.2 million per day at $75 per barrel. Production flows from up to 24 wells at once, all attached to the Mars platform. It was jointly developed by Shell (which owns 71.5%) and BP (owning 28.5%). Shell Deepwater Production, Inc. is the operator. Mars is an excellent example of modern offshore technology. It utilizes a Tension Leg Platform (TLP) that weighs about 36,000 tons. The platform is over 3,200 feet tall from the seabed to the top of the structure. It is designed to withstand hurricane force winds of 140 mph or more, and waves of 70 feet. Over 100 workers can live onboard at any one time, coming and going via helicopter, while supply ships keep a stream of necessary materials on hand. Oil production is dispatched via an underwater pipeline to Louisiana in a 116-mile trip. Gas production is sent 55 miles to a separate facility. Mars is so big that it represented about 5% of all U.S. Gulf of Mexico production when Hurricane Katrina slammed into it in 2005. A couple of days after the storm left the Gulf Coast, I had a chance to examine early aerial photos of Mars that clearly showed a disaster, something that looked like a giant collapsed factory with broken cranes and a collapsed rig tower. Nonetheless, Shell quickly repaired Mars, improved the platform's safety precautions and restored production ahead of schedule.

Despite the disaster on the Deepwater Horizon, deep offshore exploration and production is a segment of the oil industry run not by bunglers, but by very smart engineers, geophysicists and technicians. One of my friends in the business recently compared the complexity of establishing a new well in extremely deep water to the difficulty of sending astronauts to the moon. Each big, deep well represents years of highly specialized work, the effort of thousands of people, and a financial risk in the hundreds of millions of dollars.

Energy Entrepreneurs

During the later years of the previous century, entrepreneurs with stars in their eyes and a burning passion to change the world burst full force upon the Internet industry. Anything that could be shifted online was attacked with a youthful vigor that was backed by vast troves of venture capital. People throughout the world are now enjoying the fruits of that Internet boom. Today, there is another focus: since the beginning of the 21st century, entrepreneurs have been charging into the energy industry in growing numbers, with the same lust for change. They have created a cornucopia of exciting ideas for energy conservation and efficiency, such as evolutionary changes in automobiles and building systems, as well as big ideas for better energy transmission and storage. Electric generation is also high on their agenda—solar, wind, safer nuclear, cleaner coal, wave power. Then there is the stellar increase in the discovery and production of unconventional oil and gas. Taken together, it's a long and powerful list, and many of these efforts have been backed by significant sums of government subsidies and stimulus, venture capital and corporate funding. If the world is going to provide power to a growing middle class and another 2.5 billion members of the human race in a sustainable and efficient manner, then the task undertaken by these pioneers is vital to the Next Boom.

From Safer Nuclear to Ocean Wave Power, Innovators and Technology Lead the Way to More Energy Options

Long-term electric generation solutions receiving significant sums for research and commercialization include:

- Next-generation Nuclear
- Wave Power
- Hydrogen Power
- Concentrated Solar Power (CSP)
- Nanotechnology-Enhanced, Thin-Film Solar Cells

What follows is a brief discussion of some of the most intriguing technologies under research or development for electric generation. At present, all of these forms of generation suffer from high capital costs, and all require heavy government subsidies in some form or other. Nonetheless, advancing research may lead to significant improvements in efficiency and total installed costs. Advanced nuclear technology has tremendous long term potential, and it is being aggressively installed in many nations as a way to generate large amounts of power that can be counted on for 24/7 reliability. Wind energy is conspicuously absent from this discussion. While wind energy technology research is certainly underway, including efforts at the National Renewable Energy Labs in the U.S., I am not convinced of any near-term potential to dramatically improve wind generation efficiencies or reduce capital costs. Wind electric generation systems tend to operate at rates vastly below their capacities—they are only as reliable as the wind and weather. Solar power systems generally operate at much higher capacity levels than wind power, but they have historically been much more expensive to install than land-based wind systems.[19]

Nuclear: If the necessary capital can be raised, we will soon see the first new nuclear plants constructed in America in many decades. The plants will be based on next-generation technology, providing enhanced safety and longer periods of time between required maintenance cycles. However, American construction will be minor over the near term compared to that in other nations. The United Arab Emirates (UAE) has a dramatic $41 billion nuclear development plan designed to meet an electricity demand projected to double by 2020. India, which had 17 nuclear plants operating as of 2010, has firm plans to build at least six additional plants over the near term. Even the green-leaning nation of Germany may drop an existing law that requires that all 17 of its nuclear plants be shut down by 2022. (These plants provide about 23% of Germany's electricity.) On a global basis, leading industrial companies are competing fiercely to get construction contracts for new nuclear plants, with China emerging as the biggest nuclear story by far. In December 2006, Westinghouse (owned by Toshiba in Japan), a major maker of nuclear power plants, announced a multi-billion dollar deal to sell four new nuclear plants, all advanced models, to China. The deal includes work to be performed by U.S. engineering giant Shaw Group, Inc. AREVA Group, based in France, also has a deal to provide China with two reactors and approximately 20 years worth of atomic fuel.

Westinghouse, like competitor GE, is focusing on an advanced, water-cooled reactor technology. Westinghouse calls its newest unit the AP1000, and the first such plant began construction in March 2009 at Sanmen, in the Shejiang province of China. China is expected to have more than 100 reactors by 2030, up from 11 in 2009—a nine-fold increase. The AP1000 is considered to be a reactor based on generation 3+ technology. Such reactors feature higher operating efficiency, greater safety and designs that use fewer pumps and other moving parts in order to simplify construction, allow for easier operation, and make emergency responses more dependable. "Passive" safety systems are built-in that require no outside support, such as external electric power or human action, to kick in. For example, the AP1000 features systems for passive core cooling and leak containment isolation. Passive systems rely on the use of gravity, natural circulation and/or compressed gas in order to react to emergencies in a failsafe manner.

The long term future of nuclear plant design may be exemplified by the work of Babcock & Wilcox.[20] This company was founded decades ago as a manufacturer of boilers, and it has a lengthy history of making components, such as pressure vessels, for the energy industry. It has adapted its technology base in order to design an innovative nuclear generation concept, the mPower modular reactor. It is a low cost, high efficiency design of compact size. The engineers leading the project were focused on one concept: to create a reactor pressure vessel, the core of the unit, that can be built at a factory and be small enough to fit on a railroad car for delivery to the final site. This overcomes the massive, custom engineering and construction challenges that typically drive the cost of a site-built nuclear plant to more than $6 billion, and the time required for completion to several years. One modular unit from Babcock & Wilcox could be installed in relatively short order, and could power a single large industrial complex or a few thousand homes. Several of these small, low-cost units could be combined at one site to create power stations of enough overall capacity to power a small city. The air-cooled design is described as passively safe. The firm already has a lengthy history in the nuclear field, as it builds reactors for the U.S. Navy.

Several other firms are pursuing the mini-reactor business, including Westinghouse and NuScale Power. Santa Fe, New Mexico-based Hyperion Power Generation hopes to be able to sell a small reactor, suitable to power about 20,000 American homes, for only $50 million. Hyperion is utilizing technology that originated more than 50 years ago at the nearby Los Alamos Labs. Competitor Toshiba hopes to install a test unit of its "4S" (supersafe, small and simple) mini-reactor in a remote village in Alaska in the near future.

Wave Power: Eventually, the oceans could provide a small portion of the world's electricity. The main benefit of tidal power, in comparison with other forms of renewable energy, is its predictability. The timing and force of tides can be predicted with great accuracy, and thus so can the power produced by a plant. The main drawback of this source is its high initial equipment cost. Regular maintenance may also be costly. Nonetheless, throughout the coastal regions of Europe, governments are keen supporters of wave power. From the estuaries of the U.K. to the coasts of France and Portugal, tidal power is being built, tested and refined in pilot projects. For the United States, Hawaii has long been considered by some to be an ideal spot for wave power development. Several projects are under testing or consideration there, including a unit installed by Ocean Power Technologies, Inc. at the Kaneohe Marine Corps Base in February 2010.

Hydrogen: Despite the fact that fuel cell-powered vehicles are still a distant dream, hydrogen power remains a viable long term option, although it faces many obstacles. While commercial use of hydrogen fuel cells remains elusive due to high initial costs and the difficulties of distributing and storing hydrogen, research into this field is far from dead. Keep an eye on Bloom Energy's unique fuel-cell technology, currently in testing by a handful of major electricity users. Bloom claims that its patented solid oxide fuel cells use lower cost materials and can convert fuel into electricity at a much more efficient rate than other designs.

Concentrated Solar Power (CSP): As of 2010, if I was making a bet on a solar energy source, it would be on CSP, also known as "solar thermal." As opposed to solar cells, which require expensive installation on large horizontal areas, CSP uses the laws of basic science to gather and intensify sunlight, which is then focused on a given point in order to capture the resulting heat. When I was a little boy, I was fascinated by an object that was sold in a roadside gift shop during a family trip. It was a "solar cigarette lighter," a palm-sized parabolic mirror made of polished stainless steel, and I purchased one out of the meager funds in my pocket because the science of it fascinated me. A cigarette could be placed in a wire holder that rested a couple of inches above the mirror's center. When the mirror was aimed at sunlight, it concentrated the light on the cigarette's tip, thanks to the mirror's concave shape. Sure enough, my father was easily able to light a cigarette with it. This was a miniature version of the CSP technology that would fascinate me again decades later.

Today, CSP is used to heat fluids to extreme temperatures (up to 750 degrees Fahrenheit), which produces steam that then drives a turbine. The turbine powers an electric generator. A pioneer in this kind of solar power is Ausra, Inc., acquired in early 2010 by global atomic energy technology giant Areva, perhaps as a modest hedge on its giant nuclear bet. For the greatest efficiency, CSP might be combined with advanced power storage technologies. For example, systems are being tested that use CSP to heat large quantities of molten salt during daylight hours, and then use that trapped heat to create steam to turn turbines for electric generation during evening hours, steadily maintaining output until the sun rises again the next morning.

CSP can also be combined with photovoltaics. SolFocus, Inc., based in Mountain View, California, has developed solar arrays that use just one-thousandth as much semiconductor material as standard photovoltaic solar panels. The arrays are set with curved mirrors that focus sunlight onto solar cells measuring one-square centimeter, concentrating the light 500 times. The firm had raised $77 million in venture capital by mid-2009. SolFocus had installations in California, Spain and Hawaii by 2010.

There are several major CSP facilities generating electricity around the world today. Most of them were massive construction projects that required extensive engineering and investment. Typically, a CSP plant involves a large, central tower which is the focus of thousands of mirrors. These towers can be as tall as skyscrapers and can take a long time to build. A highly innovative CSP firm named eSolar may have the best idea. Its theory is similar to that which Babcock & Wilcox is applying to nuclear reactors: smaller and factory-built is better. In the same way that B&W plans to build nuclear reactor vessels that can fit on a rail car, eSolar has designed a small, two-piece CSP tower that can be built in the factory, shipped by rail and installed in one day.[21] The tower is the focus of rays gathered by a surrounding field of mirrors. The mirrors are flat and roughly the scale of a mid-sized television screen. Each mirror is mounted on a motor-driven axis that enables the mirror to follow the sun throughout the day. When solar heat from the array of mirrors hits the tower, it boils water that makes steam and turns a turbine. The turbine powers an electric generator. Best of all, eSolar's mirrors and related array frames are designed to be manufactured in a robotic factory, after which they can fit in standard freight containers for shipment by rail. Final installation at the site is relatively fast and easy. The cost savings over traditional CSP may be dramatic. In fact, an eSolar facility may prove to be considerably less expensive to build than a standard photovoltaic solar cell plant of similar capacity.

Thin Film Solar: Nanotechnology is enabling the creation of efficient solar collectors ("photovoltaic," or PV cells). Nanosolar, Inc. is a leading company in the thin-film solar field. By 2009, it had received $500 million in financing. The company's unique technology enables it to deposit nanoparticle ink onto a thin-film surface in a high speed process similar to printing. High speed production lines may eventually enable the company to achieve a relatively low investment per watt of electricity delivered. Instead of silicon, which is used in conventional PV cells, Nanosolar relies on copper-indium-gallium-selenide as a semiconducting material. A well-funded competitor, HelioVolt Corp., uses the same material. This is a significantly different technology from traditional PV, which relies on silicon materials incorporated into bulkier, more expensive units. The challenge for thin-film companies has been to enhance the efficiency of the units. Nanosolar claims that its solar cells can convert sunlight into electricity at a high rate of efficiency.

Energy in the Near Future

The near future of energy is about more efficient homes and buildings. It is about continuing progress in transportation efficiency. Like other technologies discussed in this book, it is about smaller, faster, better, cheaper. (Lee Schipper, a Senior Engineer at the Precourt Energy Efficiency Center at Stanford University, points out that air transportation in developed countries today uses 50% to 60% less energy per passenger-kilometer travelled than it did in the early 1970s, and trucking uses 10% to 25% less fuel per ton-kilometer.)[22] Energy will be enhanced by the application of technologies—ranging from biotech to information technology to nanotech—that can get more production from both new and existing oil and gas wells, and deliver more energy from renewable sources, while greatly reducing carbon emissions and the intensity of energy needed per unit of GDP. Cables made from nanotubes have the potential to enhance the world's aging electric grids. For example, electrical cables for long distance power transmission from remote solar and wind farms might prove to be most effective if made from nanotubes. The future of energy will be built upon convergence, entrepreneurship and innovation.:

If "smart grid" equipment comes to life in an economically feasible fashion, designed to allow consumers to better monitor and control energy use at the local level, while enabling electric utilities to better distribute and monitor electricity, it will be thanks to miniaturization and convergence. Over the long term, electricity from CSP and other advanced solar technologies may reach costs at or near those of electricity generated with coal or natural gas. If this occurs, then burning coal "becomes nonsensical," as Michael Moe puts it, and driving electric cars based on advanced, cost-effective batteries "is a no-brainer." However, we are still a long, long way from achieving the cost-efficiency required to bring this about.

While there is a long list of good news about the future of energy, there is also a vexing problem: misguided government subsidies. It's one thing for government to provide grants and incentives for research. It's another for government to pay producers to generate electricity that is not, and perhaps never will be, cost-competitive. This is particularly true in the U.S. and Europe, where governments are already spending more in general than they can afford to. A positive future for the energy sector lies in the application of technologies and processes that evolve to become efficient enough in operations and low enough in cost to attract conventional funding and investment. For example, if entrepreneurs bring outstanding solar technologies along to the point that they can compete head-on with electricity generated by coal, gas or nuclear fuels, then solar will gain market share. It may even be reasonable for government to support development and installation of renewable and alternative energy systems for a limited period of time if they are clearly on a near-term path to becoming cost-competitive. On the other hand, there is strong potential for special interests to convince government to sustain non-viable energy alternatives that will never be financially self-supporting. Wind power is so expensive and unreliable that it may fall into that category. Such funding could be much better spent elsewhere. Even from a carbon emissions point of view, government support can be a disaster. For example, there is strong evidence that American government support for ethanol and European subsidies for biodiesel refined from palm oil have been environmentally counterproductive.[23]

Controversies and disagreements over where, how and at what cost to produce energy will always be part of life, whether that energy comes from nuclear, oil, coal, natural gas, solar or other sources. We can count on two factors that will clearly continue to define energy over the long term. First, the demand for energy is so great and the financial impact of its production and consumption so large that competing interests will always have vast incentive to battle each other for funding, resources, markets and government favor. Next, technology and entrepreneurship will continue to create the tools and processes needed to produce the energy we need while making continuing progress in conservation and efficiency.

Internet Research Tips

Plunkett's Next Boom video for Chapter Seven:
www.plunkettresearch.com/NextBoom/Videos 🎥

Video: "Drilling for Shale Gas," MIT Technology Review www.technologyreview.com/video/?vid=439 offers an extremely interesting look at drilling and completion methods, including steps taken to protect water tables 🎥

Bloom Energy's fuel cell discussed on CBS *60 Minutes*, February 2010: www.cbsnews.com/stories/2010/02/18/60minutes/main6221135.shtml 🎥

U.S. National Renewable Energy Laboratory, basic explanations of a wide variety of renewable energy technologies:
www.nrel.gov/science_technology

Ocean Wave Power is explained at:
http://ocsenergy.anl.gov/guide/wave/index.cfm, the site of the OCS Alternative Energy and Alternate Use Programmatic EIS, U.S. Department of the Interior

eSolar CSP tower video:
www.esolar.com/our_projects/video_tour 🎥

YouTube: A Look at Life on an Offshore Oil Rig
www.youtube.com/watch?v=Hy6M0II_I1A 🎥

Join in the discussion!

- See the Reading Group Guide in the back of the book.
- Go to Facebook, search for The Next Boom.
- Join The Next Boom group on LinkedIn.

— chapter eight —

ADVENTURES IN HEALTH CARE—$175,000 WORTH OF PROTON BEAMS AND DR. SHETTY'S $2,000 HEART SURGERY

"We tend to overestimate the effect of a technology in the short run and underestimate the effect in the long run."

-Roy Amara (scientist and former president of the Institute for the Future)

Technology's Circuitous Path from the Bomb to the Hospital

One of the great ironies of science lies in the fact that the development of the most horrible weapon ever unleashed on man led to significant advances in cancer therapies that save lives every day via radiation. Radiation was a medical technology that advanced slowly over several decades. This is not surprising, as new technologies are rarely ushered from the laboratory to widespread commercialization by one visionary in one smooth process. While nuclear technology was moved dramatically forward by research that led to the first atomic bomb, radiation had, in fact, been used in cancer treatment for a few years before the Manhattan Project advanced our understanding of nuclear science.

The first noteworthy pioneer in the use of radiation to treat cancer was John Lawrence. At one time a physician on the staff of the School of Medicine at Yale University, John was the brother of Ernest Orlando Lawrence who won a Nobel Prize for his 1929 invention of the cyclotron, a circular device that emits charged subatomic particles as they are propelled through a magnetic field in a high speed spiral. John became excited about the prospect of using his brother's cyclotron to generate radioactive particles for medical use, and in 1936 he left Yale to join Ernest full-time at the Donner Laboratory in California (now known as the Lawrence Berkeley Laboratory, part of the University of California at Berkeley). Here, John made medical history when he treated a leukemia patient with a radioactive isotope, which appears to be the first time that radiation was used for the treatment of a human disease.[1]

In 1946, our second pioneer, physicist Robert Wilson, wrote a paper, *Radiological Use of Fast Protons*,[2] proposing the application of a unique type of radioactive particle for medical treatment. Wilson was a giant in the world of physics. As a young man, he was a leader in the development of the atomic bomb during the Manhattan Project, where Robert Oppenheimer appointed him head of the cyclotron group. After the bomb starred in the closing chapter of World War II, he worked at Cornell University to establish an early particle accelerator. Eventually Wilson became the founding director of Fermi National Accelerator Laboratory (Fermilab) in Batavia, Illinois, about 45 miles west of Chicago, where he oversaw the construction of a much larger particle accelerator and led the organization's extensive nuclear research, which continues today. He was acutely aware of the properties of various types of atomic particles. The ways in which they interacted with organic material, including living tissue, intrigued him. Wilson knew that medical radiation was typically based on x-rays (photons), quite different from the protons he recommended in his 1946 paper. While Robert Wilson correctly envisioned that treatment with protons might offer significant advantages to cancer patients, it was more than 40 years before a health care visionary brought this idea into everyday use in a major hospital.

In the early 1970s, a few hundred miles south of the Lawrence lab in Berkeley, a physician named James Slater, the third pioneer in our story, was well established at one of the world's most interesting hospitals, the Loma Linda University Medical Center (LLUMC). Operating in Loma Linda, California and sponsored by the Seventh-day Adventist Church, this institution is a major trauma center for the Los Angeles area, one of America's largest teaching hospitals and a major research institute. The fact that the hospital has a lengthy history of leading-edge work in many fields encourages doctors like Slater to bring bold ideas for innovative treatments to the attention of the institution's top management. For example, it was here that surgeons pioneered the earliest heart transplants for infants. Although Dr. Slater, a 1963 graduate of LLUMC's school of medicine, chose to become a physician, he also had developed a substantial interest in the field of physics and he had carefully studied Robert Wilson's theories about proton radiation.

Slater was aware that a substantial number of cancer patients had been treated with protons through the years, but only in what were essentially physics research labs inspired by Robert Wilson's theories. The first such treatment was in the Donner Laboratory in 1954.[3] On the other side of the nation, the Harvard Cyclotron Laboratory soon developed a technique for treating eye cancer with protons, with a near-perfect cure rate. This was an exceptional, vision-sparing breakthrough. Other labs, in locations ranging from Japan to Switzerland, had also conducted proton experiments with cancer patients who had tumors that sat relatively near the skin's surface, such as eye, neck and head tumors. This was an important development because, as Robert Wilson had foreseen, protons could be shaped and guided in such a way that they avoided damage to the massive numbers of exceptionally delicate nerves that lie adjacent to cancers in the head or neck. However, those laboratory settings left a lot to be desired, and the machines that were in use lacked sufficient power to treat tumors that lay deep within a patient's body. Slater was convinced that a significant advancement was called for. He knew that it would be both a daunting technical challenge and a major financial risk. Slater developed a powerful vision of a revolutionary, hospital-based proton facility at LLUMC, one that might someday offer better treatment options to patients suffering from a wide variety of cancers. However, before he could bring that vision to fruition, he had to raise tens of millions of dollars. At the same time, he needed a qualified technical partner, which he found in Robert Wilson's Fermilab, the organization that eventually designed and installed the particle accelerator and delivery system for Loma Linda.

Converging Technologies Focus on Cancer

In October 2010, a large celebration marked the 20th anniversary of the opening of Slater's proton center. More important than its hospital setting was the fact that this was the first proton unit constructed with enough power to reach cancerous tissues far within the body. This was a perfect example of technology convergence, as Slater and his team were able to design and install this clinic by merging several vital factors. One was advanced imaging. CAT scans that could enable Slater to exactly target the location of a cancer weren't available until the early 1970s, with MRIs entering the market a few years later. Another factor was advanced computer science, enabling the use of specialized software to create a radiation plan for each individual patient that involves conforming the radiation beam so that it will irradiate an exact area of the body—maximizing the chance of killing the cancer while minimizing harm to the rest of the patient. The continuing technology trend of miniaturization boosted the usefulness of computers in this regard. The largest factor was advanced nuclear science, especially the ability to create the first proton system that delivers continuously variable energy[4]—enabling the machinery to emit precisely timed doses of radiation with incredible accuracy, with the dose's output reaching its maximum at the tumor's location while leaving very little dose elsewhere.[5]

Not long ago, I climbed around the working end of this system under the guidance of a managing physicist. The accelerator weighs 50 tons. It consists of a ring of eight electromagnets of incredible power that guide and accelerate the radioactive protons. This ring of magnets measures 60 feet in circumference. As Richard Schaefer's history of LLUMC, a book titled *Legacy*, explains, "These electromagnets bend and focus the [proton] beam around a closed path within a vacuum tube. Protons are accelerated at up to half the speed of light by applying a radio-frequency voltage in synchronization with the circulating beam."[6] When the protons reach the desired speed, they are directed, on demand, to the center's treatment rooms. In these special rooms, patients are held immobile within body molds that have been custom-formed for each patient, and radiation is sent to exactly the right spot in their bodies through a complex system of imaging, software and guidance systems individually modified for the patient and constantly checked and re-checked. After observing this immense, complex system and the number of people required to plan, monitor and deliver treatment, it was easy for me to see why proton therapy is costly.

Imaging Has Elevated Radiation to a High State of Precision

The broad trends of miniaturization and convergence helped technology advance to the point that a determined doctor can build and operate his or her own nuclear facility in a local office. I visited at length with a young doctor who runs a state of the art radiation clinic on the edge of an affluent neighborhood in central Houston, Texas. Behind well-shielded walls within his building sits a particle accelerator, about the size of a small automobile. Using advanced imaging techniques, he is able to carefully design a radiation plan tailored to each patient (usually men who have prostate cancer), and can treat as many as one patient every 15 minutes with x-ray, modulated and pinpointed with great accuracy so that the cancerous area of the body receives the most significant part of the radiation dose. Since treatments require patient visits five days per week for several weeks on end, this type of small, neighborhood clinic is a huge breakthrough, and the image-guided treatment has been beautifully designed to limit unnecessary exposure to radiation. Patients receive personal attention from a small, dedicated staff, park near the door, and can be in and out in minutes after receiving state of the art treatment. This type of IMRT (intensity modulated radiation therapy) x-ray radiation is common in America today, and it represents a high level of technical and medical excellence, as does a similar technology, IGRT, which stands for image guided radiation therapy. Nonetheless, healthy tissue receives a shot of radiation. If the radiation oncologist and radiation technologist design and perform the treatment correctly, the dose outside of the targeted tumor is limited to a low to moderate amount. However, there may be side effects to surrounding tissue over a series of treatments which typically takes from 38 to 45 days to deliver in increments.

Science has struggled for decades to bring a continuing stream of small enhancements to the way that radiation is delivered, constantly tweaking technology in order to solve this problem. As Carl Rossi, a treating physician and researcher in the proton center at LLUMC once described the objective, "If one could devise a perfect radiation beam…it would deposit its entire dose in the target."[7] In its simplest use, traditional x-ray-based radiation creates a stream that you might describe as beginning with an entry dose, sending the dose through the body, and ending with an exit dose. When used to treat cancerous tumors today, modern technology shapes and conforms x-ray doses, through methods such as IMRT, so that radiation is generally concentrated on the tumor. Also, the radiation enters the body at multiple spots, spreading the entry and exit doses around.

Protons—A Step in a Different Direction

At LLUMC, Dr. Slater felt strongly that radiation should be limited as much as scientifically possible to the tumor in question. Robert Wilson's ideas intrigued him. He knew that protons are a different type of particle altogether. To begin with, they contain a positive, electrical charge, where x-rays don't. This charge gives them unique physical properties. The discovery of a physics phenomenon called the Bragg Peak enables very exacting release of proton radiation in the desired spot.[8] By carefully timing the radiation beam while controlling its acceleration, the charge can be designed in such a way that the dose will enter the body, moving at one-half the speed of light, but nonetheless remain nearly benign until it suddenly "peaks" at the site of the tumor at a power of up to 250 million electron volts.

Deep underground at LLUMC's proton center lies a large, wood-paneled waiting room. In groupings of sofas and arm chairs you are likely to find a few women awaiting treatment for breast cancer, and one or two men or women who are awaiting their daily dose targeted at a brain tumor, neck tumor, eye cancer or lung cancer. At a game table you might see a handful of supportive spouses assembling a large jigsaw puzzle. However, the majority of the seats are filled with men who are undergoing treatment for a male curse: prostate cancer. The prostate is surrounded by extremely fragile nerves that serve, among other things, erectile function and bladder control. If these nerves are damaged during treatment, the patient could end up impotent, incontinent, or both. Cancer of the prostate is a serious disease. If the patient has an aggressive form, it can quickly spread to the rest of the body. There are many options for treatment, so many that it is extremely confusing to most patients. Surgical removal of the prostate is a common choice, often through the use of an exacting da Vinci brand robotic surgical device. X-ray radiation is another common choice. Many men get through the surgery or standard radiation without lasting harm. A significant portion of the patients, however, endure side effects that last from several months to the rest of their lives, during which time they end up wearing a diaper or living without a normal sex life.

The Physician's Preferred Medicine

On my first tour of LLUMC, I was escorted around the proton center by a patient who had that day completed the last of his 44 treatments for prostate cancer. This patient, a man who had spent his life working in various technology sectors, and his wife of only a few months had cancelled their honeymoon when he was diagnosed with prostate cancer, and they got him registered as a patient for proton therapy as quickly as possible. As we talked over coffee, they both exuded a combination of confidence, joy and relief. He was cured; he felt great; he was going home as a fully functioning man. In other words, Dr. Slater's convergence of technologies had delivered on its promise.

As we toured the facility, my new friend introduced me to many of his fellow patients. They had travelled from all over the world to be treated here. Two patients I met that day were physicians themselves. Later I met a colo-rectal surgeon from the Midwest who had chosen to be treated for prostate cancer here. "I've performed surgery on too many men who needed me to straighten out complications from traditional prostate cancer treatments," he said. "I wouldn't be anywhere but right here for my therapy." In a Loma Linda townhouse development that is often home to patients from out of town who need a place to stay during their 9 to 10 weeks of treatment, I talked at length with the wife of a man who was likewise being treated for prostate cancer. "He's an oncologist himself," she explained. "Several of the doctors in his practice group back home travelled here to be treated before him." Obviously, a significant number of physicians are enamored of the exactness of proton radiation when it comes to their own bodies.

A special section of the waiting room in the LLUMC proton center contains toys, games, children's books and a TV set that is usually playing a Disney video. At first glance, I thought it was a place where the children of patients could keep themselves occupied. In fact, the room was for young cancer patients, some of them barely old enough to walk. Their best chance for survival and a return to healthy growth might reasonably be based on a radiation method that targets their tumors while causing the least possible peripheral damage to a young, growing body. A small number of parents who understand these advantages bring their children to LLUMC's proton center. The children receive a mild anesthetic prior to entering the treatment room to eliminate any chance that they will move during the radiation session, which lasts only a few minutes.

Why Isn't Every Cancer Patient Treated with Protons?

Clearly, there was something exciting going on here in Loma Linda. So exciting, in fact, that, as of mid-2010, a total of nine proton radiation centers were operating in America, including a relatively new site at the MD Anderson Cancer Center in Houston as well as a University of Florida center. Two more were under construction in the U.S., and an additional two were in planning. Several more proton centers are scattered around Europe and Asia.

So what's the problem? In a word: money. Proton therapy is frowned upon by some observers who claim that any added health benefit is not worth the increased cost. A modern proton treatment facility requires an investment in the neighborhood of $200 million. This will create a clinic with five treatment rooms, offices, waiting rooms, and most costly of all, the immense accelerator and the software and instrumentation required to operate it in a way that will support a unique radiation plan exactly suited to the needs of each patient. The list price for a patient's prostate cancer treatment in such a proton facility in the U.S. is about $175,000. (In the rare cases that an insurance provider covers it, or when it is covered by Medicare, the negotiated price may be considerably less, which is typical with all types of medical treatment.) In comparison, treatment in an x-ray-based, state of the art IMRT clinic costs closer to $40,000, and the investment required to set up such a clinic might be in the $4 to $6 million range for a modern space housing one treatment room, a small accelerator and advanced imaging equipment. Eventually, the high price of constructing modern proton radiation centers and then using them to treat patients at great expense led to a controversy that rages on today.

There is No More Effective Way to Learn about the Health System than to be Sick Yourself

In early 2007, I was diagnosed with a fairly aggressive prostate cancer. Multiple doctors, including an internist in my family, told me I absolutely should not delay treatment beyond a few months. My personal experience well illustrates both the challenges and the opportunities of the health care sector during the Next Boom. While exploring the wide variety of potential treatments (surgery, standard radiation, radioactive seed implants and high intensity focused ultrasound or HIFU—which is not yet available in the United States, but something I could have received in Canada), I learned that there are no guarantees of a return to health. Many men die—about 258,000 yearly worldwide. I watched two men in my family suffer and eventually die from prostate cancer: a father-in-law and a step-father. It was ugly. I became acutely aware that many patients are injured by the usual cures. I had no interest in dying a lingering, painful death from cancer. I was also not at all interested in giving up sex or wearing a diaper. This was personal, and I decided to go to war against the disease, utilizing every possible skill as an analyst that I had acquired over the years, and every possible personal connection I could mine.

I spoke with some of the most highly regarded experts in the realm of cancer treatment, including department heads at the famous MD Anderson Cancer Center in Houston. One of Anderson's doctors tried to put me at ease about side effects by explaining that the hospital had a complete "penile therapy" department, where pumps and other implants might be surgically deposited in a non-functioning organ. Needless to say, this didn't make me feel any better. After three months of intense research, interviewing doctors, visiting hospitals, enlisting the aid of three different pathologists, and enduring, at various highly regarded clinics and hospitals, multiple physical exams, CAT scans, MRIs, blood samples, diagnostic agent injections, a biopsy, x-rays and waiting room delays, I chose to be treated by Slater's machine at LLUMC.

Taking the Plunge—Zapped by 250 Million Electron Volts Daily

I rode the elevator down into LLUMC's proton treatment area several days a week over a period of 10 weeks. Once I had reached the bowels of the building, which are heavily shielded by thick concrete and steel against radiation leakage, I would sit in the waiting room, anticipating a "we're ready for Mr. Plunkett" call from a technician while I contemplated the strange twist of genetic fate and my stubborn research that had combined to bring me here. Once called, I would put on a hospital gown, climb into my personal plastic body mold and assume a supine position, from which I stared up at a proton beam-carrying gantry weighing several tons, looming over me like a great white heron holding breathlessly still before it plunges into the water to strike a fish. The technicians would buzz about, expertly prepping my body for treatment, turning cranks, checking meters and screens. When my body alignment was painstakingly adjusted using imaging and computers, a physician double-checked everything. Next, a technician would speak into an intercom that connected him to the accelerator control room, "we're ready for the beam in room three." The crew then ducked behind heavily shielded walls. I waited alone for the protons to pour forth, essentially immobilized, immodestly garbed by the gown. In my mind, I imagined what was going on behind the antiseptic white walls, wondered whether I would be cured and wondered how this disease and its complex, costly treatment would change me or my outlook on life. Sometimes there were system delays of several minutes that left me with my reveries. Other times I waited only a brief time before I was zapped for a couple of minutes. Once the beam started, I knew it was there thanks to moving machinery and the whir of motors, but I didn't feel anything. The next morning, I would be back for more. As a result, my cancer appears to be completely gone, and I have endured no loss of body function—I am as normal as I wanted to be when I started on this path.

How Much is the Best Cure Worth to the Growing Global Middle Class?

We now have the ability to apply a near-endless array of technology to patient care and the treatment of disease. The problem, of course, is the ballooning price tag. Overall, my own cure cost a small fortune, well over $200,000 for doctors' visits, various procedures and tests, imaging, radiation and travel. It was my choice to take this route, and I am fully aware that most people wouldn't do the extensive research I did, and couldn't pack up and move to Loma Linda, California for nearly three months of treatment as my wife and I did, shipping personal goods, three computers and a mountain of business equipment and files so we could continue to work. I am also aware that the cost of such a plan is prohibitive. In one vivid picture, that is the dilemma of modern health care: who should receive what treatment, to what extent, at what cost, and who should pay for it? While this topic is already a major controversy, as seen in the acrimonious debate leading up to the recent passage of a major health care bill in the U.S., that controversy is about to move to a new level.

As the global population expands by nearly 2.5 billion people over the next four decades, as hundreds of millions of people enter the middle class for the first time, and as the populations of the most advanced economies skew dramatically toward old age, health care will undergo a significant evolution. Increased demand will lead to more research and innovation. Increased demand will also lead to an ever greater need to create cost efficiencies and faster, cheaper paths to the best possible outcomes. For example, entrepreneurs are hard at work on ideas that will greatly reduce the cost of proton therapy. One theory for advancement comes from Still River Systems, based in Littleton, Massachusetts, which raised $33 million in venture capital in 2009. Working in partnership with MIT's Plasma Science and Fusion Center, this firm hopes to build smaller facilities containing only one treatment room, intended to be much simpler to design and install, and much less expensive at a cost of $20 million or so. This is the smaller, faster, better, cheaper concept again, a theme that will recur more and more often in the health sector during the Next Boom.

Another breakthrough in proton therapy was announced at a 2008 meeting of the American Association of Physicists. Thomas Mackie of the University of Wisconsin is developing a machine that uses a dielectric-wall accelerator (DWA) in a unit that is much smaller than standard systems. Rather than speed up particles over a long distance, a DWA speeds particles through massive electric fields over distances of little more than two yards. If successful, the new machine also has the potential to reduce the cost of a proton system significantly, and the size of the accelerator system from several hundred square feet to a device that fits into a small room. Meanwhile, at LLUMC and elsewhere, highly automated treatment rooms are being installed that greatly increase the number of patients that proton centers are able to handle each day, thus amortizing a clinic's capital cost over a much wider base of patients and potentially reducing treatment costs.

Health Care Delivery Will Soar on a Global Basis

In terms of cost and scope, the next leg of global health care delivery will be one of the largest strategic, financial and moral challenges ever faced by humanity. Who receives the $100,000 surgeries? The $400,000 intensive care hospital stays? The $10,000 per month specialty drugs? Better still, how do we drive down the costs?

A comprehensive study published by the OECD[9] in 2010, covering more than 30 nations including the majority of the world's most developed economies (but not Brazil, Russia, India or China), found stark contrasts between health costs in the United States and those of other countries. In 2008 (the latest complete data available), a list of nations that includes, for example, the U.K., France, Germany, Mexico, Canada, South Korea, Japan, Australia and the U.S., spent an average of 9.0% of GDP on health care. The highest figures were in America at 16.0% of GDP, France at 11.2% and Switzerland at 10.7%. Health expenditures per capita, on a purchasing power-adjusted basis (PPP), averaged $3,000.

The numbers were skewed significantly by the United States, where per capita health costs are vastly higher than in other nations. It remains to be seen what the net effect of the 2010 health care act in America will be, but there is little doubt that expenditures could rise rapidly for the near term, as a result of the legislation and the needs of aging Baby Boomers. Health care as a percent of GDP in the U.S. was an estimated 17.3% in 2010 and could approach 20% by 2020 unless significant cost controls are established.[10]

Total health care expenditures around the world are difficult to determine, but $5.5 trillion would be a fair estimate for 2010. That would place health care at about 8% of global GDP, with expenditures per capita about $800. This $5.5 trillion breaks down to approximately $2.6 trillion in the U.S., $2.4 trillion in non-U.S. OECD nations and $0.5 trillion elsewhere around the world. Outside the U.S. and the rest of the OECD, that would allow roughly $88 per capita per year. Clearly, there is vast disparity in the availability and cost of care among nations, as there is with personal income and GDP. Health care spending per capita in the U.S. was equal to about $8,289 in 2010, while spending in the world's remotest villages was next to nothing. The trend over the near future is for the modest amount now spent on health care in emerging nations to rise dramatically, while OECD nations like America struggle to contain their own mountainous costs.

The coming steady increase in the amount spent on health care in emerging markets is going to be a gold mine for innovators and entrepreneurs. A study published by Yes Bank and The Associated Chambers of Commerce and Industry of India forecast that health spending in India will continue its current rapid growth, reaching $77 billion, or about $70 per capita by 2012. Analysts at McKinsey estimate that the number of people in India with some sort of health insurance stood at about 100 million in 2009, and will climb to 220 million over the mid-term.[11]

China is in the process of implementing a broad, national boost to health care availability. In 2009, the Chinese government announced a plan to cover 90% of its population with some sort of basic health insurance by 2011. However, health facilities and services available will remain relatively limited for some time, particularly in the remote villages and towns. China's more ambitious plan is to have an affordable health services system available throughout the nation by 2020.[12] In other words, China intends to introduce relatively modern health care to more than 1 billion consumers for the first time. This will be an immense undertaking in a nation with few doctors who have been trained to Western standards. Novartis AG, one of the world's largest drug manufacturers, sees so much potential in the growing Chinese market that it announced it will soon increase its Shanghai R&D center from 160 employees to 1,000. By 2014, the firm expects China to be its third-largest market.[13] Experts at Bayer AG, a leading global chemicals and drug manufacturer, expect the Chinese health care market to explode over the near term, growing from $25 billion in 2009 ($19 per capita) to $80 billion in 2013 and $220 billion by 2020 (about $169 per capita).[14] However, even after this massive increase, health care will account for only about 2% of the Chinese economy. In order to provide advanced health care, such as diagnostics and drugs, at such low cost, a vast amount of innovation and creativity will be required. This is also true in India. Fortunately, the best of this innovation will eventually spill over into the developed world.[15] While initial steps into modern health care in China and India were based on equipment, procedures and technologies from the West, health care delivery in the future will accelerate quickly based on locally developed efficiencies and technologies.

Cutting Technology Costs to the Bone

With the same spirit that Bharti Airtel has slashed the cost of providing cell phone services to Indian consumers, and Tata has used every possible engineering twist to create the $2,500 Nano automobile, engineers in India are building advanced medical diagnostic equipment that can be sold at prices dramatically lower than their Western equivalents. Designers at GE Healthcare-India realized that new levels of innovation had to be achieved to meet the unique challenges of the growing health market in that nation. To begin with, of course, there's the basic fact that there is much less money available for patient care than in the West. If engineers can dramatically lower the capital cost for equipment, then modern health care technology can spread, based on local financial means. There are other challenges as well, however. Electric power is less reliable and not as widespread as in the West, which means that battery-powered equipment is often preferable. Lighter weight, portable equipment is also desirable, which can enable medical technicians to travel with their machines from village to village.

GE's Logiq 100 portable ultrasound device (a basic imaging tool used for a wide variety of diagnostic purposes), developed initially for the Indian market, is compact and inexpensive.[16] It weighs only 10 kilograms, compared to as much as 200 kilograms for a traditional machine, while costing about 60% less. In addition to serving the Indian market, GE is now able to sell this machine in many other emerging nations. Similarly, the Mac 400 battery-powered portable electrocardiogram (ECG or, sometimes EKG, an important diagnostic tool used on a noninvasive basis to check heart conditions) is also a significant breakthrough for markets like India. In some clinics, a patient can receive an EKG test for the equivalent of $1, thanks in part to GE's Mac 400.[17] Eventually, a breakthrough like this will benefit the U.S. and other mature economies in their struggle to lower medical costs. An enhanced, but still low-cost, model called Mac 800 has been developed by GE for the U.S. market. Chinese medical equipment manufacturers are also aggressively entering the global market at very competitive prices. This will likely become analogous to the telecommunications equipment market, where Chinese manufacturers like Huawei quickly captured market share in this multi-billion dollar sector on a worldwide basis, sometimes crushing legacy Western companies that charged much more for similar goods. For example, Jiangsu-based orthopedic implant maker China Kanghui has sales offices in 24 nations. In particular, it is targeting parts of Europe, plus emerging markets in South America and Africa. It went public in 2010, and its stock is listed on the NYSE.

Global Competition is About to Set the Health Care Industry on Fire—Dr. Shetty's $2,000 Heart Surgery

Of course, the cost of equipment isn't the only challenge faced by health care providers. The strategies and practices used to deliver care to the patient are paramount. Innovative clinics and hospitals in the emerging world have the potential to turn the health care delivery business upside down. One of the world's most closely watched doctors is India's Devi Shetty, a surgeon trained in London who has become an extremely successful medical entrepreneur. While he gained fame at one time as Mother Teresa's heart surgeon, he is best known today as a builder of highly cost-effective hospitals. Shetty has been compared to America's Henry Ford who, in one of the most important innovations in industrial history, launched the modern automobile manufacturing industry by creating an assembly line to turn out a high volume of quality cars at affordable cost. Before Ford, the car was just a distant, unaffordable dream to most people, and was assembled slowly, in low volume at great cost—somewhat like the way health care is delivered today. Dr. Shetty, in a revolutionary manner reminiscent of modern industrial practices, employs the economies of scale offered by high volume output to deliver quality surgery at extremely modest prices.

In the United States, open heart surgery can cost $100,000, although negotiated insurance agreements generally make the price much less. Costs of this magnitude clearly won't work in India, as that $100,000 is more than the lifetime income of an average couple at today's levels. At Shetty's massive, 1,000-bed privately-owned hospital in Bangalore (as much as five times larger than a typical American hospital), open heart surgery costs only about $2,000,[18] and the outcomes are excellent. The hospital is like a surgery factory. Dozens of operating theaters give this center the capacity to perform up to 70 heart surgeries in one day.[19]

There's more. Adjacent to his Narayana Hrudayalaya Hospital for heart care, he has built a 1,400-bed cancer center and a 300-bed facility for eye care. Another Shetty unit provides a broad range of dental care; yet another runs a stem cell bank to enable mothers to deposit their babies' umbilical cords for potential use in the future for stem cell therapy. Additional surgical specialties at Shetty's hospitals include neurology and orthopedics. He recently raised millions of dollars with the goal of building several more medical centers in India, hoping for 30,000 total beds.[20] Shetty's hospital is also a partner in a "micro health insurance" program called Yeshaswini, which it states covers nearly 3 million farmers at a monthly premium of 10 rupees each (about 25 cents). In addition, Shetty hopes to build a 2,000-bed acute care hospital in the Cayman Islands, where an easy flight to this Caribbean haven could lure U.S. patients. This is a logical extension of his business, as "medical tourism" is one of the fastest-growing trends in the health care industry. Millions of Western patients now travel yearly to medical centers in Singapore, Thailand, India and other spots where they can receive treatment at bargain rates from doctors who were educated at some of the world's best teaching hospitals in Europe and America.

Biotechnology—About to Break Through to a Higher Level

The Holy Grail of 21st Century medicine is to know the presence or absence of key genes (or gene mutations) within individual patients, and then tailor each patient's treatment according to their genetic makeup. This vision includes knowing that a patient will react positively to a given drug, due to the presence of a specific genetic mutation. It also includes knowing that a patient will likely have a negative reaction to certain drugs, thus allowing doctors to avoid treatments that would create unwanted side effects. When this goal is achieved to a broad degree, outcomes could improve dramatically, billions of dollars currently spent on ineffective treatments yearly could be saved, and harmful side effects could be dramatically reduced.[21] This is not an impossible dream. Simply put, the biotech era is about to come into full bloom. A convergence of advancing computer technology, decades of intense research and a greater understanding of how drugs interact with the unique DNA of individual patients is finally beginning to ramp up individually targeted "personalized" medicine. This trend is moving slowly at first, not only because the challenges involved have proven to be more daunting than first imagined, but also because research and clinical trials take a great deal of time and money. Nonetheless, it will unleash a flood of better, more targeted drug strategies over the long term. This will be an immense leap forward from the traditional practice of internal medicine where, in an educated guess, physicians prescribe drugs that appear to be the most hopeful for a given condition, on blind faith that the patient's body will have a positive reaction. While this approach often works, it also frequently fails due to currently unknown reasons.

The first patients to benefit significantly from genetically targeted drug delivery were women suffering from breast cancer. A drug often prescribed today, based on individual genetic conditions, is Herceptin, a monoclonal antibody that was developed by Genentech. Approved by the FDA in 1998, Herceptin, when used in conjunction with chemotherapy, shows great effectiveness in controlling breast cancer for certain patients who are known to "overexpress" the HER2 protein (that is, there is an excess of HER2-related protein on tumor cell surfaces, or there is an excess presence of the HER2 gene itself). About one-quarter of breast cancer patients have tumor cells that overexpress HER2. Such cells tend to grow quickly and are more likely to lead to a recurrence of cancer. (A test, the immunohistochemistry or "IHC" assay, can be used to determine whether overexpression exists in the patient's tumor cells. In some cases, a test referred to as fluorescence in situ hybridization or "FISH," is used to determine whether there are excess HER2 genes per chromosome.)

Personal DNA Maps—Starting as Low as the Price of a Bicycle

A useful genetic test is marketed by Genomic Health, based in Redwood City, California. Its Oncotype DX test provides breast cancer patients with an assessment of the likelihood of the recurrence of their cancer based on the expression of 21 different genes in a tumor, including HER2. By mid-2010, more than 90,000 patients had used the test. Such tests can assist breast cancer patients in evaluating the results they may expect from therapies, such as chemotherapy, or from the use of Tamoxifen, a hormonal drug therapy that appears to block the growth of breast cancer cells. Genomic Health is also doing full-scale clinical development on a test to predict the likelihood of a recurrence of colon cancer. Such tests will be standard tools of treatment for cancer and other diseases in coming decades. Meanwhile, now that personal DNA testing is available at relatively modest cost, the federal government is trying to determine how to best regulate such tests and how to insure that consumers are receiving accurate information from test providers.

By 2010, as many as 1,400 genetic tests were available from various suppliers, including 23andMe's use of microarrays (a lab on a chip used to analyze genetic samples) that test human saliva for up to 1 million genetic variations linked to disease (about $429) and Navigenics' similar test that detects predisposition to diseases such as Alzheimer's and type 2 diabetes ($999). Higher-priced services, costing up to $100,000, come from companies like Knome that analyze an individual's entire genome and provide genetic variation data that can't be detected by microarrays. Such services represent a huge drop in cost from the $3 billion in public funds spent by the Human Genome Project, and the significant sums invested by private company Celera Genomics, to complete the first maps of the human genome in the early 2000s. Prices may fall even more thanks to a $10 million award offered by the X Prize Foundation to the first scientific team to sequence 100 complete human genomes in 10 days for less than $10,000 each. Likely contenders for the prize include Illumina, Inc.; Life Technologies, Inc.; 454 Life Sciences (owned by Roche Holding AG) ; and Helicos BioSciences Corp. Another startup called Pacific Biosciences hopes to sequence a genome in 15 minutes for less than $1,000 by 2013.

The scientific community's continuously improving knowledge of genes and the role they play in disease is leading to several different tracks for improved treatment results. One track is to profile a patient's genetic makeup to determine whether certain defective genes are present and are causing an illness. Yet another application of gene testing is to study how a patient's liver is able to metabolize medication, which could help significantly when deciding upon proper drug dosage. The use of specific medications based on a patient's genetic profile could greatly boost treatment results while cutting costs. A widely quoted 1998 study published in the *Journal of the American Medical Association* estimated that 2.2 million Americans suffer serious, adverse side effects from prescription drugs yearly. Of those, an estimated 100,000 die, making adverse drug reaction (ADR) a leading cause of death in the U.S.[22]

Gene Therapy

Another track under research is to attack, and attempt to alter, specific defective genes—this approach is sometimes referred to as "gene therapy." Generally, pure gene therapy would target defective genes within a patient by introducing new copies of normal genes. These new genes can be delivered through the use of "vectors," such as viruses or proteins that carry them into the patient's body. Since 1989, more than 1,400 clinical trials of some form of gene therapy have been conducted, with only 47 reaching phase III trials and none making it to regulatory approval in the U.S. as of early 2010. The challenges are daunting. Attempts to develop a useful gene therapy may fail because they are rejected by the patient's immune system, the vectors may create unwanted side effects, or the therapy, once injected into the patient, may be so short-lived as to be of little use. As you will see, a convergence of biotechnology with nanotechnology (for drug delivery) may be the key to effective gene therapies.

China-based Shenzhen Sibiono GeneTech achieved the world's first pure gene therapy to be approved (in China) for wide commercial use, in October 2003. The drug is sold under the brand name Gendicine as a treatment for a head and neck cancer known as squamous cell carcinoma (HNSC). The company that developed the drug claims that it has proven effective in the treatment of several types of cancer during trials; however, observers in other nations are highly skeptical and want to see more extensive data. The drug is not available outside of China.

Meanwhile, applications of gene therapy are undergoing research in the U.S. and elsewhere for treatment of a wide variety of diseases. For example, gene therapy may be highly effective in the treatment of rare immune system disorders, melanoma and cystic fibrosis. In 2006, two men suffering from a rare malady called chronic granulomatous disease (CGD), which makes patients terribly vulnerable to infections, were able to cease taking daily antibiotics due to a gene therapy that introduced healthy genes to replace defective ones in their bloodstreams.[23] In 2008, a team of scientists led by a group at the University of Pennsylvania successfully introduced a healthy gene to six patients suffering from Leber's congenital amaurosis, a condition that leads to blindness. Four of the patients' vision improved. It is small victories like this that keep researchers energized and research dollars flowing. Gene therapy is still in its infancy, and is not without its failures. However, over the very long term, the potential for gene therapy and genetic testing used for the choice and dosage of medications may be tremendous.

A Genetic Map Saves a Baby's Life

A report in 2009 in the *Proceedings of the National Academies of the Sciences* related how a genomic map had altered the treatment of a patient: an infant in Turkey thought to have a possibly fatal kidney disease called Bartter's syndrome. The baby was constantly dehydrated and was unable to properly gain weight. The baby's doctors, seeking confirmation of this rare disease, sent a blood sample to geneticist Richard P. Lifton at Yale Medical School. Lifton ran a limited genomic study of the baby, focusing on an important DNA group known as the exome, and found genetic mutations that indicated the problem was something else entirely, a condition known as "congenital chloride diarrhea."[24] This meant that the physicians were able to treat the child correctly, in an early instance of a genomic map leading to a significant change in the treatment of an extremely ill patient.

Nanotech Will Boost Health Care

Nanotechnology is being applied to advances in imaging, diagnostics and bone replacement, and that list will expand continually. We are quickly learning how extensively nanotechnology and biotechnology can support each other. The result over the long term will be a revolution in the way that diseases are diagnosed and the methods with which drugs are delivered. Because of their small size, nanoscale particles can readily interact with biomolecules on both the surface and the inside of cells. By gaining access to many areas of the body, nanoparticles have the potential to detect disease and deliver treatment in unique ways. Nanotechnology will create "smart drugs" that are more targeted and have fewer side effects than traditional drugs. When combined with a growing understanding of how to personalize medicine based on individual genetic makeup, the results may be revolutionary. Eventually, nanotechnology-biotechnology convergence will lead to faster cures and lower costs, but we still have a long way to go.

Current applications of nanotechnology in health care include drug delivery technologies for immunosuppressants, hormone therapies, drugs for cholesterol control and drugs for appetite enhancement. For example, the NanoCrystal technology developed by Elan, a major biotechnology company, enhances drug delivery by reducing otherwise large drug particles into tiny bits, typically less than 2,000 nanometers in diameter. By reducing the particle size, the drug's exposed surface area is increased, solving a key drug industry problem by making compounds more soluble, so they will have greater uptake by the patient's cells. In the language of the drug industry, the goal is to increase "bioavailability," and the NanoCrystal has been able to increase this measure by a factor of up to 600%. Nanotech can be used to provide more effective delivery of drugs in tablet form, capsules, powders and liquid dispersions. Abbott Laboratories uses Elan's technology to improve results in its cholesterol drug Tricor. Par Pharmaceutical Companies uses NanoCrystal in its Megace ES drug for the improvement of appetite in people with anorexia.

Elsewhere, researchers are learning how to deliver gene therapy using nanoparticles. At the School of Pharmacy in London, doctors successfully wrapped genes in nanoparticles in such a way that the gene had no effect until it was in the presence of a cancerous environment. Once triggered, the gene stimulated production of proteins that destroy cancer.[25] This could lead to an alternative to chemotherapy and its numerous toxic side effects. Magnetic nanoparticles are under development at the MIT-Harvard Center of Cancer Nanotechnology Excellence that can detect as few as two cancer cells in a small biospecimen in minutes. In a test on mice, this breakthrough detected tumors with nearly 100% accuracy.

At the University of Michigan at Ann Arbor, Dr. James Baker is working with molecules known as dendrimers to create new cancer diagnostics and therapies, thanks to grants from the National Institutes of Health and other funds. This is part of a major effort at the Michigan Nanotechnology Institute for Medicine and Biological Sciences. A dendrimer is a spherical molecule of minute size (five to 100 nanometers) and well-defined chemical structure. Dr. Baker's lab is able to build a nanodevice with four or five attached dendrimers. To deliver cancer-fighting drugs directly to cancer cells, Dr. Baker loads some dendrimers on the device with folic acid, while loading other dendrimers with drugs that fight cancer. Since folic acid is a vitamin, many proteins in the body will bind with it, including the proteins on cancer cells. When a cancer cell binds to and absorbs the folic acid on the nanodevice, it also absorbs the anticancer drug. For use in diagnostics, Dr. Baker is able to load a dendrimer with molecules that are visible to an MRI. When the dendrimer, due to the folic acid, binds with a cancer cell, the location of that cancer is shown on the MRI. These nanodevices may eventually be developed to the point that they are able to perform several advanced functions at once, including cancer cell recognition, drug delivery, diagnosis and reporting of cancer cell death.

In summary, the convergence of nanotechnology with biotechnology has the potential to radically change cancer therapy for the better and dramatically increase the number of effective therapeutic agents. Nanoparticles can serve as customizable, targeted drug delivery vehicles capable of ferrying doses of drugs or therapeutic genes into malignant cells while sparing healthy cells, greatly reducing or eliminating the often unpalatable side effects that accompany many current cancer therapies.

Health Care Costs in the U.S.

In the wake of the tremendous growth of all aspects of the health care industry from the end of World War II onward, efficiency, competition and productivity were, regretfully, largely overlooked. Much of this occurred because employers, plus federal and state governments, paid such a large portion of health care bills. Insured patients virtually never asked their doctors or hospitals about prices, and no advance information was offered by the providers. Why should patients care what it costs if they know they don't have to pay the bill?

Physicians are caught between their desire to provide quality care and the desire for cost control on the part of payors, including insurers, Medicare and Medicaid. An energetic cost versus care debate existed for many years prior to the immense American health care act of 2010. Big questions loom over the health care system like a dark storm cloud. Should an insurance company pay for a patient's proton beam therapy, or should it force the patient to elect standard radiation or surgery at one-fourth the cost? To what extent should patients be responsible for the costs of care and the results of their own health habits? What is the return on investment of health education and wellness programs? What expenses should society be willing to bear to briefly extend the life of a chronically ill, aging patient?

A study released by the Milken Institute in 2007 found that 109 million Americans suffered from one or more of the most common chronic diseases, including cancer, diabetes, heart disease, pulmonary conditions, mental disorders, stroke or hypertension.[26] This meant that more than one-third of all Americans had these conditions to one degree or another. The study estimated one year's cost of treatment of these illnesses at $277 billion, but estimated lost economic productivity due to illness to be vastly higher at $1 trillion. (The study was based on 2003 data.) In other words, lost work and lost output caused by chronic diseases, much of it preventable by better health habits, reduced the nation's GDP by as much as 10%. These effects might be vastly reduced through consumer health incentives and preventive medicine. For example, obesity, lack of exercise and cigarette smoking are immense contributors to chronic diseases. The Centers for Disease Control and Prevention reported that medical costs for obesity-related diseases alone rose to $147 billion in 2008, compared to $74 billion in 1998.

Then there is the debate over health care "reform." Passions were running so high on both sides of the debate in America during 2009 that fights broke out at town meetings held by members of Congress while visiting their districts. Many citizens were venting their anger. They were either strongly opposed to an expansion of government control of health care or they were strongly in favor of universal coverage with extensive government management and funding. Of those opposed, many saw further government involvement as a step toward unwanted socialism, arbitrary health care rationing, big increases in spending and wasteful, bureaucratic programs. Of those in favor, many saw the government as the only hope for coverage for all, regardless of their station in life, employment or personal income. Arguments at these meetings often became so intense that name calling and shoving ensued. *The New York Times* reported in September 2009 that a man in Ventura County, California had part of his little finger bitten off in a fist fight at one of these meetings. Fortunately for him, his post-fight emergency room treatment was covered by a government program, Medicare.

Out of dire financial necessity, sometime in the future a societal focus on the costs of problems such as obesity and smoking will eventually lead to better health education and wellness incentives, and preventive medical techniques will evolve that save lives and money. With the cost of American health care galloping toward 20% of GDP, concrete steps are needed to put an end to the most wasteful leaks in the health care basket. Here are a few ideas for lowering the cost of care in the U.S.:

1) Rein in the costs of malpractice claims—approximately $75 billion[27] in yearly costs (2010 estimate) due to malpractice insurance costs, lawsuits and defensive medical practices aimed at lessening the risk of being sued. The exact causes and costs are a controversial subject, but several states, including Texas, have already passed very effective caps on malpractice lawsuit damages.
2) Put Americans on a diet and exercise plan—$160 billion[28] in yearly costs (2010 estimate) of obesity-related diseases.
3) Stop fraud, abuse and billing errors in health insurance claims, Medicare and Medicaid—an estimated $68 billion yearly (2007).[29] Better claims processing technology that utilizes artificial intelligence and outsources processing to firms with significant financial incentives (if better results are achieved)

might be a powerful combination.

4) Cut administration costs, speed the flow of patient and claims data, create better management practices and reduce mistakes in patient care (such mistakes are an immense expense—electronic patients records could be a big help if properly applied, and adding significant layers of artificial intelligence would be even better). An industry-wide focus on operating efficiencies and the establishment of "best management practices" could bring significant results. For example, a 10% reduction in annual costs at doctors' offices and clinics would result in $54 billion in savings and a similar reduction in costs at hospitals would result in $79 billion in savings (2009).[30]
5) Reverse innovation—the health industry should be willing to adopt proven, effective, cost-saving techniques, technologies and practices as they evolve in emerging nations.
6) Make it a national priority to promote wellness, health education and disease prevention. Japan, which provides a national health plan to residents and has very small numbers of obese citizens, is requiring waist measurements for men and women ages 40 through 74 as part of annual checkups. Men's waists must measure less than 33.5 inches. The maximum is 35.4 inches for women. Those who exceed the limits receive dietary guidance and monitoring until excess weight is lost.

The Biggest Challenge

If creativity and innovation are going to enable cost-effective delivery of health care to a growing, global middle class totaling billions of people, then the engineers, entrepreneurs, scientists and doctors doing the work are going to need a significant boost from evolving technologies. Fortunately, the timing of this global health care expansion is excellent, as a convergence of multiple fields of science and technology is already occurring that will speed progress and potentially reduce the costs of care.

An important foundation in life sciences has already been laid. High speed computers, as well as enhanced database and statistical software, have converged with chemistry and biotechnology. This enabled such advances as lightning-fast analysis of potential drugs, major genomics projects, microarrays and the launch of new biotech drugs. As 2010 began, this convergence was gaining significant traction, and beginning to lead to advances in personalized medicine. In fact, converging information technologies will lead to better health care in general, thanks to the collection of more patient data, deeper analysis of the treatment outcomes of millions of patients at once, and, eventually, better patient health management.

While technology and innovation are going to revolutionize the way we deliver health care and cure disease, common sense is also required. If we fail to deter obesity and smoking, if we fail to make people responsible to a reasonable extent for their own habits, health choices and costs of care, then we might win the battle against diseases like cancer, but lose the war against ballooning costs.

Internet Research Tips

Plunkett Video for Chapter Eight:
www.plunkettresearch.com/NextBoom/Videos

Dr. Shetty's Narayana Hospitals in India:
www.narayanahospitals.com

Loma Linda University Medical Center's Proton Radiation Unit:
www.protons.com

From Fermilab's official history pages, a description of the advantages of proton radiation and Fermilab's role in developing the equipment at Loma Linda:
http://history.fnal.gov/Neutron_Therapy/index.html#fermilab

Inexpensive genetic profiles? Pacific Biosciences genetic sequencing videos,
www.pacificbiosciences.com/aboutus/video-gallery

The Alliance for Nanotechnology in Cancer (part of the National Cancer Institute), offers terrific videos and slides explaining the potential of nanotechnology as a cancer fighting tool. www.nano.cancer.gov

Join in the discussion!

- See the Reading Group Guide in the back of the book.
- Go to Facebook, search for The Next Boom.
- Join The Next Boom group on LinkedIn.

— part four —

CHANGE

— the next boom —

— chapter nine —

EDUCATION, GENERATIONS AND CHANGE

"To expect the unexpected shows a thoroughly modern intellect."
-Oscar Wilde

Some of the most important changes of the near future are going to be generational—including a significant tide of people entering the later stages of life while, on the younger end, more than 90 million members of Generation Y mature in the U.S. It is vital to analyze these generations if you want to have an understanding of future trends and coming changes, and this chapter does so in descending order, starting with the elderly and working through to youngsters. Later in this chapter is a discussion of looming changes in the education of the youngest generations and the importance of education to the Next Boom. Although this discussion will be about Americans and their futures as students, consumers, workers and retirees, it also contains implications for many other nations, particularly those with large Baby Boomer or school age segments.

Change: The Older Generations

The Elderly, Boomers and Those In-between—Working Longer and Worrying More

If 60 is the new middle age, as some observers claim, let's say that the elderly are people who have inhabited Planet Earth for 75+ years. Journalist Tom Brokaw referred to Americans in this age group as the "Greatest Generation" because they successfully endured the challenges of World War II and many of them grew up during, and despite, the Great Depression. Later they saved their money and worked hard to help get America back on its feet. There is no denying that they have seen a lot. In the U.S., the elderly totaled an estimated 18 million in mid-2009, including 5.5 million who were 85+ years in age.

Today's seniors are solid evidence of increasing life spans. Because of their advanced ages, the elderly account for a large proportion of national health care expenses. Virtually all of them qualify for Medicare or VA (Veterans Administration) care, and millions who have low incomes rely on Medicaid to pay for nursing home expenses or other extended care. According to the Health Services Research and Educational Trust, over 60% of nursing home care is paid for by Medicaid, a program that was designed to cover low-income people, including seniors and children.

The elderly who are fortunate enough to own much in the way of assets found that, as of 2008-2010, their portfolios were down and the value of their homes, often paid for, had declined. In any event, they don't shop as much as they did when they were younger. Many are in good physical condition, mentally sharp, and active in businesses or careers, volunteerism, traveling, family activities or all of the above. As a group, however, their financial habits and consumption are not going to have a major impact on the Next Boom. The most significant economic effect they will have on the near future is the public burden of providing their health care. As you will see, a much more significant impact will come from a massive group of people who are marching steadily down the path toward old age, but have not yet arrived at its doors: Baby Boomers.

Meanwhile, there is definitely a discernable generational slice between Baby Boomers and the elderly. This group is often called Pre-Boomers. They were born between 1935 and 1945. As of 2009, that made them 64 to 74 years old. In the past, Pre-Boomers often quipped, "I'm not going to leave anything to my children; I'm going to spend it all first." After the onset of the Great Recession, however, they are likely to be much more financially conservative. They will probably live a long time, and they are going to cost the health care system a fortune. In the U.S., this generation totaled about 23 million as of mid-2009. These people are experienced and street smart. They watched their own aging parents go through the later stages of life. They have observed the debilitating effects of Alzheimer's disease, arthritis and dementia, and they don't want to be poor, sick or incapacitated when they are elderly themselves. They want unlimited access to the best possible health care. Many Pre-Boomers on the younger end of the scale still work. Many of the rest of them are going back to work, because they got crunched by the big bust.

Baby Boomers Approach Their Senior Years

The term Baby Boomer was coined to describe a group that includes the children of soldiers and civilian workers who were involved in World War II. When those veterans and workers returned to more normal post-war lives, they made babies, lots of them, born from 1946 to 1964. As a result, the Baby Boom generation is one of the largest demographic segments in the U.S., making up about 25% of the population. These people originally numbered about 80 million, and they totaled about 76 million survivors in 2009. In 2011, millions will begin turning traditional retirement age (65), but I predict that normal retirement age will soon be thought of as 70, and the benchmark age for "early retirement" will be 65. Eventually these hordes of Baby Boomers will result in major, long-term growth in the elderly portion of the population.

As of 2009, you could be as young as 45 or as old as 63 and be a Baby Boomer (at this point I should disclose that I am one of them). Baby Boomers will have an immense effect on the near future and the Next Boom. For example, they will be a big drain on Social Security. There will be continuing debate over whether and how to reform America's Social Security system, and that debate will accelerate now that Boomers are becoming eligible for benefits. When the Social Security Act was passed in 1935, the typical American was much younger than today (the median age in 1930 was only 26.5 years compared to 35.3 years in the Census of 2000), and the ratio of workers to retirees in the U.S. was about 42 to 1, thanks in part to vast numbers of recent

Estimated American Population by Generation, 2009

Generation	Estimated Population
The Elderly	18 million (born 1934 or earlier, ages 75+)
Pre-Boomers	23 million (born 1935-1945, ages 64-74)
Baby Boomers	76 million (born 1946-1964, ages 45-63)
Generation X	66 million (born 1965-1980, ages 29-44)
Generation Y	91 million (born 1981-2002, ages 7-28)
Diversity Generation	32 million (born 2003+, ages 6 or less)
Total	306 million

Note: Some analysts use slightly different dates to describe Generation Y.

Source: Plunkett Research, Ltd. estimates based on U.S. Bureau of the Census data.

immigrants. By 1960, the ratio of workers paying into the Social Security system to retirees taking out was a relatively comfortable 5 to 1.[1] In 2010, there were only three workers contributing to Social Security per pensioner. In 2020, the ratio will be only about 2.6 workers per retiree,[2] a fiscal nightmare. That projection for 2020 assumes there are no changes in the way the system collects taxes and pays benefits, but changes will occur in the near future as a fiscal necessity. Future retirees will be restrained from receiving benefits in some manner, whether by income tests or later retirement ages, while workers and their employers may find themselves paying higher tax rates.

In 1935, when Social Security was established, the average life expectancy of a 65 year-old man was a short 12 years. The cost of providing him with modest Social Security benefits until death was relatively low compared to the cost of providing generous benefits to a 65 year-old man today for the rest of his life, which will likely average 19 years in the near future. In 1930, there were only 6.6 million Americans age 65 and above; in 2010 there were about 40 million, and that number will increase dramatically as the Baby Boom generation hits retirement, to the point that it will double to 80 million between 2010 and 2040.

If you study the history of the Franklin D. Roosevelt administration, you see that FDR intended Social Security to be a conservative, self-funding program that operated as a true trust account, where tax proceeds would be banked to pay for future retirement needs. Unfortunately, the U.S. government has a history of looking for ways to spend more money today while delaying the pain of repayment until tomorrow, and the Social Security piggy bank has been hard to resist.

For decades, the federal government has utilized excess Social Security taxes somewhat like current income, while booking a phantom "trust account" amount, leaving it to future politicians to worry about how to pay future Social Security benefits. That is, surplus Social Security taxes have been lent to the U.S. Treasury and used to fund other needs. Eventually, either benefits must be cut or the Treasury will have to repay the Social Security system by issuing debt or finding new tax revenues. The bill that is soon coming due is immense, and the problem will be compounded astronomically by the costs of the Medicare and Medicaid systems.

Baby Boomers know that they are likely to live for a long time, and many want solid investments and financial planning that will enable them to live well. An average American female aged 60 can expect to enjoy about 25 more years of life. Unfortunately, some Baby Boomers on the older end of the scale feel crushed by circumstances. They aren't old, but they aren't as young as they used to be. Many had been planning to enjoy affluent retirements, based on their stock portfolios and retirement plans, plus what they thought would be the ever-rising values of their homes. Now, post-Great Recession, many are planning to work until they are 70 or 75. When asked, many will tell you in earnest that they are never going to retire. They are suddenly saving money with a focus they did not have in the past. They don't want to be old and broke.

The newly enhanced enthusiasm for thrift and savings by this massive slice of the population has positive implications for the Next Boom. Future, accumulated savings of Boomers will be put to good use in business and industry in the form of investments in stocks and bonds, and will help keep interest rates in check. Eventually, however, their savings and investment balances will decline as they draw down funds needed to sustain them in their later years.

Older Workers and the Next Boom

Keeping Baby Boomers in the workforce for an extended period could have very positive effects on the economy of the near future. In 1983, when I began studying employment trends and data while writing my first book, *The Almanac of American Employers*, a new edition of which is still published yearly, I developed a theory that Americans of the future would be working much later into their lives than their parents had. Because of several factors, including longer life spans, better health in their older years, rising costs of living, the decline of defined benefit pension plans, evolving technologies and increased global business competition, I believed that people would be re-careering multiple times, and many would be going back to school on and off to train for future career changes. Today, my thoughts of 1983 are truer than ever. For example, as a result of the long-term decline in employment opportunities in manufacturing, many adults have been enrolling in school to retrain as nurses, or to learn additional computer skills so they can launch new careers. Baby Boomers who work into their 70s will be a vital part of the workforce, and they will contribute significantly to GDP growth. A 2010 study by Anne Shattuck, published by the Carsey Institute at the University of New Hampshire, found the number of older Americans working for pay has increased dramatically in recent years. In 2009, 22% of men over age 65 were in the labor force, along with 13% of women. Fourteen years earlier, in 1995, the numbers were only 17% and 9%.[3] In addition, the number of seniors who work full-time, as opposed to part-time, is rising significantly.

By the early 2000s, many employers were already developing human resources strategies aimed at hiring or retaining older workers. On the lower end of the scale, retailers like Home Depot, a firm that has been known to need tens of thousands of new hires yearly, have found older people to be ideal employees. They have knowledge that is extremely useful for providing advice and service to shoppers. They are experienced workers who understand the need to show up on time. Wal-Mart reported in 2005 that it had 220,000 employees who were age 55 or older.[4] This trend is powerful enough that the AARP has advised members who are seeking work to consider a select list of senior-friendly employers that includes Pitney Bowes, MetLife, Principal Financial and Walgreens.

On the higher end of the employment scale, older workers with long-term experience in scientific, technical and engineering tasks will be vital in keeping the gears of business and industry turning. For example, during the 2000s boom, when the airline industry saw good growth, rules were altered in the U.S. to enable commercial airline pilots to keep flying until age 65, instead of facing forced retirement at age 60 as they had in the past. Heavy industry is an even better example. The National Science Foundation estimated that 38% of America's scientists and engineers were 50+ years old as of 2006.[5] These older workers may not be good at creating the latest electronic games, but they are undeniably the people who have the experience and the knowledge necessary to successfully operate the laboratories, factories, refineries and legacy systems that keep industry humming. They are going to be extremely difficult to replace in America and certain other nations, partly because younger generations have been shunning degrees in science and engineering for softer disciplines. Energy firms like ExxonMobil, Halliburton and Schlumberger, and chemicals firms like global giant BASF, are dealing with two specific challenges in this regard: First, how to document and pass along the immense treasure of work-related knowledge that these older employees have, and second, how to keep these employees interested in working later into their lives. BASF, a firm with about $90 billion in annual revenues and an employee base of 90,000 scattered around the world, estimates that by 2020, 50% or more of its employees will be 50 to 65 years old. It has implemented measures ranging from making the workplace more comfortable and safe for older workers to creating an intense knowledge transfer program where older workers mentor younger staff members.

In Japan, where employers and government alike are acutely aware of the challenges posed by a dramatically aging population, many companies are anxious to hang on to older workers. Pensions are often modest in Japan, so older people are motivated to remain in the workforce. Some employers find it expedient to re-hire workers after retirement, putting them in lower pay scales with simpler responsibilities. The older workers remain active and receive income, while their stress and duties at work are lessened.

Currently, incentives are built into the U.S. tax and pension structure that encourage people to retire. There is the Social Security system, which enables retirees to begin drawing full benefits at age 66 if they were born from 1943-1954. The age requirement increases by a few months, according to date of birth, until it becomes age 67 for people born in 1962 or later. Tax laws require that people begin taking withdrawals from their tax-deferred IRAs by age 70½. What if you are still working and don't need to draw from your IRA? Considering the dire straits of the Social Security and Medicare systems, the sooner regulations change, the better. A lessening of disincentives for older people to work would relieve a great deal of the future strain on Social Security. Former Secretary of State George P. Shultz co-authored a book in 2008 with Stanford University economist John B. Shoven. In *Putting Our House in Order: A Guide to Social Security and*

Health Care Reform, they estimated that if reforms were made to the Social Security and income tax systems that would encourage older people to stay in the workforce, then an additional 10% per year would be added to America's total labor hours worked by 2050, laying the foundation for gains in GDP that might reach $1 trillion yearly in terms of 2008 dollars.[6] In the near future, legislation may start to move this way, if Congress understands that such reform would boost economic output. Older workers are part of the Next Boom.

The table on page 201 provides a look at the future—the changes that will occur in generations from 2009 through 2025. Note how a vast number of Baby Boomers will still be young enough (under 75 years) to work full- or part-time, boosting the economy through their extremely valuable work experience and knowledge. In addition, note how Generations X and Y will be in their prime working years.

Change: The Younger Generations

Generation X

Generation X is a loosely defined and variously used term that describes people born between approximately 1965 and 1980, although the date range varies from one analysis to another. For our purposes, as of 2009 they ranged from 29 to 44 years of age and totaled approximately 66 million. For several years, Generation X has been referred to as a group influential in defining tastes in consumer goods, fashions, entertainment, politics and technology. This is a generation that slowly went digital, beginning with their first Apple or Commodore personal computers in the 1970s and 1980s that were primitive compared to today's consumer electronics. Many are the leading edge and early advocates of LOHAS, lifestyles of health and sustainability.

More importantly, they are the center of the workforce of the near future and the Next Boom. Many are well educated, and at the workplace they are already providing a large number of consumers with health care, financial advice and technical services. Those with the minds and nerves for it are the big guns and hot traders in the investment world. Others are teachers who are laying the educational foundations of the younger generations. Many have gone into law. Some are architects, designers, engineers, scientists, planners and consultants who are putting in long hours at the office, or they wear the uniforms of America's armed forces. A handful will be the newest members of

Estimated American Population, 2025, by Generation, With Percent Change from 2009

Generation	Percent Change from 2009
The Elderly	30 million (born 1950 or earlier, ages 75+); +66%
Pre-Boomers	None, all will be in the "Elderly" category; -100%
Baby Boomers, who are not yet elderly*	60 million (born 1951-1964, ages 61-74); -26%
Generation X	67 million (born 1965-1980, ages 45-60); +1%
Generation Y	93 million (born 1981-2002, ages 23-44); +2%
Diversity Generation	81 million (born 2003-2020, ages 5-22); +153%
Diversity+ Generation	25 million (born 2021-2025, ages 4 or less); N/A
Total	356 million; +16%

*Baby Boomers who have not yet hit "Elderly" status. Boomers who are 75+ are counted in The Elderly for the purposes of this table. Technically, they are still Baby Boomers

Source: Plunkett Research, Ltd. estimates based on projections made by the U.S. Census Bureau in 2008. Numbers include immigration flow.

Congress. They saw the boom, and they've survived the bust, although many had their homes foreclosed upon or lost their jobs. They have been in debt for immense amounts of credit card balances and mortgages, and some are paying off the last segments of their student loans. This generation knows that job security is scarce, and that they are likely to live for a long, long time. They are tired of being in financial arrears, and they are changing their attitudes about savings and consumption. To a growing extent, their money will be going to savings, not to the mall.

Generation Xers are globetrotters and citizens of the world. Many have no qualms about hopping an airliner to India or South Africa or Australia. This generation is aware on a basic, intrinsic level that there is more to the world than their own towns and neighborhoods. They are connected digitally every minute of the day with online friends, and can chat, text, network and Skype with people important to them no matter where they are located in the world at the moment. They are citizen reporters online, posting comments, photos, reviews and videos in an endless stream. They can't hide, and you can't hide anything from them. Many members of Generation X are entrepreneurs, and they are good at it.

I have three sons who are Generation Xers. All three grew up in Texas, but two now live in Asia after completing part of their college educations there. They range from self-sufficient to affluent. They were all significantly moved by the Great Recession. One of them saw it coming and benefitted well. Another, who earns what most people would consider to be a very high salary, has a wife and family and enjoys enviable job security due to his skills and track record, had the following reaction to the recent bust: "We've quit spending," he told me in early 2009. "We're not using the credit cards any more. We're watching every expense." While I'm listening attentively to this report, I'm well aware that he makes plenty of money and will probably continue to do so for the rest of his life. Having enough money isn't the issue. "We're tired of spending," he explained. This is the new financial face of consumers who are thirty-something and forty-something.

Generation Y, The "Millenials"

Born between approximately 1981 and 2002 and numbering about 91 million in 2009, this is a group that will effect great changes over the long run. By 2025, they will be a massive presence, 23 to 44 years of age, in their prime working years. The older members of Generation Y, including today's college students and recent grads, have been bashed with a hard dose of reality—they've learned that times aren't always good, and jobs aren't always easy to find. Business cycles remain a fundamental part of life, but this was quite a revelation to young people who were either not yet born or not yet adults during previous recessions, such as those of 1990-91 or 1973-75.

They are also known as Echo Boomers, Millennials or the Millennial Generation. Because they are the leading edge, members of Generation Y are critical to the Next Boom. As a group, they ranged in age from 7 to 28 as of 2009. Consequently, the younger members of this generation make up the bulk of today's school students. The big issue is whether the education system is up to the challenge of preparing them for the future, a future that will be increasingly crowded, complex and technology-based.

Many members of Generation Y are the offspring of Baby Boomers. Perhaps their parents have gone out of their way to make certain that their Gen Y children didn't act in the manner that many older Boomers were noted for in their youth, which might be described as rebellion for rebellion's sake. In fact, many of the parents of Generation Y grew up under unique theories of parenting. Boomers were often raised according to standards laid down in extremely popular books written by Dr. Benjamin Spock—books that advocated a great deal of tolerance on the part of parents, and a great deal of freedom of self expression for their children. As teenagers and young adults during the 1960s and early 1970s, many older members of the Boomer generation burned their draft cards, used frightening amounts of recreational drugs, resisted authority or were involved in infamous bouts of civil unrest that rocked the nation, such as the demonstrations outside of the Democratic National Convention at Chicago in 1968, or the antiwar protests at Kent State University of Ohio in 1970 that resulted in the killing of four students and the wounding of several others. On the other hand, many of the younger members of the Baby Boom took a hard swing to the right, wore conservative clothes, practiced good manners and became known as "Preppies" for their mild prep school-like clothing and demeanor. Perhaps their parents had seen more than enough of rebellion by the time these younger Boomers were coming of age and raised their children to be more conservative in manner.

The most influential book regarding Generation Y is probably *Millennials Rising*, by Neil Howe and William Strauss. The authors are extremely optimistic about this group, pointing out that members of Generation Y have excellent attributes that will make them good, productive citizens. The enthusiasm that Howe and Strauss express about the Millennials has been widely noted. Early in their book, they describe these youngsters as, "More numerous, more affluent, better educated, and more ethnically diverse,"[7] than previous generations. The authors forecast an encouraging outlook, "They are beginning to manifest a wide array of positive social habits that older Americans no longer associate with youth, including a new focus on teamwork, achievement, modesty and good conduct. Only a few years from now, this can-do youth revolution will overwhelm the cynics and pessimists."[8] If they are right, maybe the next heroes, instead of the next Bernie Madoff, will come from the Millennial Generation.

The Diversity Generation, a Dawning Group

As of 2009, this generation was aged six and under, and totaled about 32 million. I call this emerging group the "Diversity Generation." It is the newest group of children, those who were, or will be, born from 2003 through 2020, who are entering the world at the very moment that I am writing this paragraph. This is the generation whose massive numbers will consist largely of people with ethnic backgrounds. The parents of many of these youngsters will be Hispanic Americans, Asian Americans and African Americans. Tens of millions of these children will be new immigrants or the offspring of relatively recent immigrants. Eventually, they will total more than 81 million. They will be the generation that benefits most from charter schools and the coming technology-based revolution in education. LOHAS and always-on pervasive computing will be basic facts of life to them.

Change: A Revolution in Education

Vast Improvements in Education Will Be Part of the Next Boom

For the Diversity Generation and the younger end of Generation Y, the quality and effectiveness of the education they receive is going to be of paramount importance to their ability to create a positive impact on the Next Boom. If you have been following the state of public lower education in the U.S., then you know it is in miserable condition. The 2009 *National Assessment of Educational Progress* found that 25% of 8th-graders could not read at a basic level (defined as "partial mastery of knowledge and skills") and only 32% achieved proficient levels ("solid academic performance") or above. Looked at inversely, a terrifyingly high 68% failed to deliver "solid performance" in reading. A glance at the scores in mathematics for 8th-graders is equally distressing, where 27% failed to perform math at the basic level, and only 34% were at or above the proficient level for their grade.[9] This is a horrible failure on the part of American society and government.

Fortunately, there are several bright lights of hope for education that will slowly but surely enable more and more young Americans to get the foundations that they need and deserve. There is strong evidence that the American public has reached its limit of tolerance with the education system's status quo. This evidence ranges from the instant popularity of the bluntly critical 2010 documentary about public education, *Waiting for "Superman,"* to the rapid rise of charter schools, to intense news coverage of the one-time head of Washington D.C.'s school system, Michelle Rhee, who valiantly fought for teacher accountability and positive change.

Immense Costs—Humiliatingly Poor Results

Before we delve into the bright spots, let's look at the vital statistics and the costs of America's public education system. There were 57.4 million students enrolled in grades K-12 in the United States in the 2008-2009 school year, including 49.8 million in public schools. They were taught by 3.7 million teachers. The annual cost of educating those children was estimated by the U.S. Department of Education (DOE) at $667 billion for schools of all types, including public, private and home-schooled.[10] (To put this cost in perspective, that $667 billion is equal to about two-thirds of the amount Americans spent on groceries at supermarkets, discount stores and retail food outlets of all types in 2009.)[11]

The DOE estimates that spending per pupil enrolled in public schools averaged $10,384 in 2008-2009 (ranging from a low of $5,706 in Utah to a high of $16,163 in New Jersey). Public schools in America are paid for by taxes, including income taxes and property taxes. These taxes are paid by wage earners, homeowners, landlords and businesses. If you are a renter, your landlord has built the cost of school taxes into your rent. Consequently, virtually every member of the resident population participates in paying for schooling, regardless of whether they have children of their own, regardless of age, regardless of whether they are working or retired. Federal dollars paid for 7.8% of public schooling in 2008-2009. This is money passed down to the states, provided that their education systems meet certain standards and follow federal guidelines. The public student body in 2008-2009 was comprised of "57% White, 20% Hispanic, 17% Black, 5% Asian/Pacific Islander and 1% American Indian/Alaska Native," to use the DOE's terminology. Those ratios are about to skew dramatically toward the Hispanic side. Unfortunately, the agency reported that about 10% of students had "limited English proficiency."

A study of the educational achievement level of 15 year-olds in 2006 among the OECD nations, including Japan, Canada, the U.K. and Germany, found America near the bottom. It ranked 25th in math and 24th in science. This is nothing short of humiliating. What is the bottom line in terms of America's economy? A McKinsey & Company study, *The Economic Impact of the Achievement Gap in America's Schools*, estimated that, had America raised its educational performance to the level of competitive nations like South Korea during the 1983 through 1998 period, then by 2008 the U.S. economy would have enjoyed a GDP somewhere between 9% and 16% greater than actual results. In other words, America would have been more innovative and competitive, household income and wealth would have been higher and there would have been many more jobs created. The same study showed what it called a "racial achievement gap," the dire need to improve the results of schooling among children of all races. McKinsey reports that, if achievement at school for black and Latino students had caught up with that of white students by 1998, then ten years later in 2008 GDP would have been between 2% and 4% higher.[12]

America's poor results from public education certainly aren't due to a lack of spending. The OECD found that for 2005, in terms of percent of GDP, education spending in the United States was equal to the average for all OECD nations, 4.8%. In addition, America spent more per student on a PPP basis than all but five OECD nations.[13]

A Tipping Point for Education Reform

The Obama administration announced that the DOE will use several billion dollars in new federal aid, in a program dubbed "Race to the Top," to encourage, among other things, states receiving these funds to have legislation in place that will not prohibit, or effectively inhibit, increasing the number of charter schools in the state or the number of students enrolled in such schools. This is part of a larger education payout from 2009's federal stimulus package, with substantial new money hoped to push the nation toward common education achievement standards on the K-12 end, and to provide more money for community colleges as well as Pell Grants for college students' expenses. To a large extent, reforms are aimed at making teachers and administrators more accountable for their results. There has been a fierce, long term battle between those who seek to improve school results in this manner, and the well-entrenched unions that represent educators.

The problems within public education are so enormous, the bureaucracy so inefficient, the issues so complicated and the teachers' unions so powerful and effective at blocking reform that the challenges faced by America are titanic. (Milton and Rose Friedman's phrase, used as a title to one of their books, "Tyranny of the Status Quo," is an appropriate description for the present K-12 education morass.) Of course, there are many excellent schools and hundreds of thousands of passionate, dedicated, highly effective teachers and school administrators in America. What we need is a lot more of them, and we need to create incentives and reform school bureaucracies in a way that will foster an evolved educational system that is up to the difficult challenges of the near future. This is not, by any means, an impossible task if Americans are willing to take up the challenge and demand that they get what they are already paying for.

I believe we are reaching a tipping point where technology and alternative schools will begin to greatly improve the education of Generation Y and the Diversity Generation. This is of paramount importance if these generations are going to be equipped to work with and advance future technologies. Revolutionary changes in American lower education will be focused on two areas: 1) charter schools or schools that adopt charter-like reforms, and 2) an increased use of technology as a boost to learning, both in and out of the classroom.

Knowledge is Power—Charter Schools Break the Mold

Charter schools, that is, unique schools that have been granted special status by legislators to operate in a non-standard manner, are hamstrung by regulations in some states and soaring ahead in others. The point of a charter school is to set up an innovative, highly efficient, highly effective school that is not bogged down by the bureaucracy and ineffective methodology often found in major public school systems. Frequently, teachers at charter schools are not members of teachers' unions, but that is not always the case. Charter school founders are mavericks who are looking to break the status quo in order to elevate learning and improve the future. These schools are a relatively new development. They have created considerable controversy, and many early schools did not work out well. Today, however, charter schools are a big trend, a beacon of hope to parents and potentially a big breakthrough in educational quality. Twenty-one new charter schools were slated to open in New York City alone in the fall of 2009.[14] By the 2009-2010 school year, according to the National Alliance for Public Charter Schools (NAPCS), such schools numbered 4,936 in the U.S., and accounted for 5.1% of all public schools.[15] NAPCS estimated that 365,000 children were on waiting lists for charter schools, partly because many states had placed strict caps on the number of charter schools, the number of students who could enroll in them, or both.

The most admired charter school system that operates on a large scale may be the Knowledge is Power Program (KIPP). Other well-established systems include Uncommon Schools, Achievement First, Green Dot and Aspire.[16] I met one of KIPP's founders, Mike Feinberg, and listened to him extol the basics upon which KIPP operates. It is common sense in action: a commitment from teachers (who must communicate personally with parents on a regular basis and be available after hours, by phone or otherwise, to assist with homework questions), longer school hours, and a commitment by students to apply themselves. Highly successful charter schools such as KIPP operate on a low- to no-tolerance of slackers basis. There are no excuses for poor performance by students or teachers, and there are high expectations of parents. These schools often focus on helping lower-income children. Since KIPP's modest 1994 founding, 82 KIPP campuses in 19 states and the District of Columbia had opened by mid-2010, enrolling 21,000 students. (A few are not officially operating as "charter" schools due to restrictive local education laws.) The organization states that 90% of KIPP students are African American or Latino, and 80% are eligible for the federal free and reduced-price meals program. Students are accepted regardless of prior academic record, conduct record or socioeconomic background. In other words, these aren't pampered middle or upper class kids from families with long histories of high achievement. The results at KIPP are astonishing. The program serves students from pre-K through the twelfth grade. In 2010, KIPP reported that, "Nationally, more than 90% of KIPP middle school students have gone on to college-preparatory high schools, and over 85% of KIPP alumni have gone on to college." KIPP is a roaring success and a prime example of America's potential to reform education, empower teachers and build better students, if Americans have the will to apply themselves to the task and demand such improvements from their federal, state and local governments. With schools like KIPP, everyone wins: students, parents, teachers and America as a whole.

Online Learning and Computers in the World's Classrooms

Campus-wide Internet access, digitization of libraries, pervasive use of computers by students and courses offered online are already revolutionizing education at the college level. The most telling statistic may be the long-term rapid growth of the University of Phoenix and the other schools owned by Phoenix-based Apollo Group, Inc. While courses are also taught in person by this firm, it is largely an institution of online learning, and its enrollment by 2009 had grown to 420,000 students. Meanwhile, online learning has become mainstream throughout much of higher education. In *Online Nation: Five Years of Growth in Online Learning,* authors I. E. Allen and J. Seaman found that 19.8% (3.48 million) of college and vocational school students throughout America took at least one college course online in 2006.[17]

In *Liberating Learning*, authors Terry M. Moe, a senior fellow of the Hoover Institution at Stanford University, and John E. Chubb, a visiting fellow at the same institution and cofounder of the private education firm EdisonLearning, show how online learning aids have a significant opportunity to elevate both the teaching profession and learning itself over a time frame of perhaps 20 to 25 years. They offer multiple examples of successful, in-classroom teaching aids that enable students to learn at their individual levels while making teachers more efficient.

India, a nation enjoying one of the world's biggest bases of computer programming professionals while facing one of the world's largest education challenges, is benefitting from an innovative online learning system called SmartClass, which is utilized by 6 million students in nearly 14,000 schools.[18] SmartClass, developed by Educomp Solutions, Ltd., delivers standardized, well-designed lessons to flat screen monitors in classrooms, while facilitators who are managing classrooms in person move the lessons along. A Smart Assessment tool is also used in some Indian schools, which delivers real-time measures of student learning based on their responses to questions when students use wireless devices to interact with the system. Educomp's professional development program has trained nearly one million teachers in technology integration and best practices. Delivering education via technology is a logical way to help India overcome teacher shortages and slim budgets in a nation of more than 1 billion people. Similar tactics are proving effective in America and elsewhere. In the U.S., online learning tends to be extremely advanced and well funded. *Liberating Learning* focuses on the fact that teacher quality is a basic requirement for quality education. The authors explain how technology, when properly applied by an enlightened administration, can lower costs and staffing needs while attracting and empowering high quality teachers. For example, "If students make more use of technology, the number of teachers in a school can be reduced significantly. If elementary students spend but one hour a day learning electronically, certified staff could be reduced by a sixth. At the middle school level, two hours a day with computers would reduce staff requirements by a third."[19] They go on to assert that the quality of teachers would be enhanced because school systems could use the resulting financial savings to increase teacher pay and training, thus attracting much better employees and doing a better job of supporting them.

Technology revolutionized American business and industry, boosting productivity, increasing quality, and enhancing information flow in a manner that created better transparency, accountability and teamwork; it has the potential to do the same in education. Online learning is quickly gaining momentum in primary and secondary education, both for home schooling and in the classroom. Many states already have well-funded virtual (online-only) schools, some of which are designated as charter schools. Leading systems include the Florida Virtual School (FLVS), the ACCESS Distance Learning system in Alabama and the Virtual Public School in North Carolina. The National Alliance for Public Charter Schools reports there were 217 virtual charter schools in the U.S. in the 2009-2010 school year, plus 132 hybrid virtual/brick-and-mortar charter schools. Learning and education support delivered via technology will make up a vital building block of the Next Boom. In his exceptional history of U.S. education reform, *Saving Schools*, Harvard's Paul E. Peterson states, "...choice and accountability, if coupled to technology, have the potential to create a more productive educational system. Elementary and secondary education cannot turn the excellence corner, so long as the industry remains labor-intensive."[20] FLVS achieved 2009 operating costs that were about 25% lower per pupil than the average school statewide.[21] By 2010, FLVS offered more than 100 courses, including 14 advanced placement courses, taught by over 1,200 staff members. For the 2008-2009 school year, 154,125 "course completions" equaling one-half credit each had been achieved by its students, nearly triple the number of 2005-2006.

Change: New Generations and Technologies Will Break Down the Status Quo, Whether or Not You Are Ready for It

If you are going to be prepared for the Next Boom, you must accept the fact that massive changes are in store. A dramatic generational evolution is going to occur. That's evolution (significant, steady changes), not revolution. In the U.S., the evolution will be fueled by three factors. 1) Baby Boomers are aging to the point that they will eventually become an immense generation of elderly people with their own unique needs. It will take monumental effort on the part of business and government to successfully accommodate those needs, and that effort, along with associated costs, will have a deep effect on Americans of all ages. 2) Younger generations are so massive in size that their energy, needs, desires, tastes, habits, ethics, talents and scientific discoveries will reshape the world to a large extent, changing the status quo in realms as diverse as entertainment, government, transportation and retailing. 3) Evolving technologies will continue to alter the way in which we work, travel, learn, shop, invest, access health care, access education, and receive news and entertainment. These technological advancements will be readily adopted by members of the younger generations. In a word, things are going to "change."

The concept of change as an immutable force existed long before we were bludgeoned with the word during the presidential election campaign of 2008. Change is continuous, but it is so easy to take the status quo for granted that many people fail to grasp this simple idea and therefore are surprised, challenged or even ruined by unexpected changes that they might otherwise have seen coming. Famed management guru Peter Drucker was known for emphasizing the need for leaders to anticipate change and design solutions in advance of changing market conditions. Simply put, the failure to anticipate and plan for change is a path toward failure.

One Point of View on Change: "Life is a Moving Target"

My professional life is focused on analyzing the trends that bring about significant changes. I fully remember the day I became alert to change as a basic concept. I was spending a week at a massive lake house in the Midwest owned by Joseph Sugarman, a bright, hyper-energized marketing guy who made a small fortune by convincing consumers that change was good for them. Joe is a man who speaks in rapid, articulate sentences. He is a unique entrepreneur who has written several books on advertising and marketing, and he once billed himself as one of the world's highest-paid advertising copy writers. Sugarman understands the power of words like Mozart understood the power of music, and he is not bashful about using that power to achieve a desired effect. In the 1970s, Joe wrote wordy, but brilliant, magazine ads that were the written equivalent of today's TV shows on the Home Shopping Network. He figured out how to make consumers desire to buy expensive, technologically advanced things that they had never heard of before. In this manner, he successfully sold the first electronic calculators, primitive machines by today's standards that were made by Texas Instruments with a price of $400 each. (Today, thanks to the technology revolution and globalization, you can buy similar calculators for $5 or less.) In other words, he convinced consumers that the change to digital technology from pencil and paper was worth the high cost because it would enhance their lives. He sold a wealth of other leading edge, early technology items, despite their high initial prices, including one of the earliest personal computers. Reading his ads in such magazines as *Scientific American* and *U.S. News & World Report* was like walking into a Best Buy store today and test-driving the merchandise. He could literally make you feel the benefits of owning the item and experience the difference it could make, even though you weren't physically in its presence. He sold you benefits. He sold you a concept. He sold you change. He should have been a politician.

Our gathering at Joe Sugarman's house made up a small, fascinating group of people interested in effective marketing strategies and better copy, including brilliant, tireless Fred Pryor, whose Fred Pryor Seminars became one of the most successful for-profit firms ever to offer seminars on business topics. Others at the gathering included a top executive from Rodale (a highly regarded publishing house that started out by printing information on gardening and grew to be the publisher of magazines like *Men's Health* and *Prevention)*, along with a man who was a political powerhouse and a famous fundraiser for conservative political candidates. Sugarman started our discussion one morning with the following, simple thought: "Change is inevitable." I now realize that this is a well-recognized idea, but I was about 26 years old at the time, and it was the first time anyone had drilled that thought home with me. Heretofore, I had been much more interested in the present than in the future. There was a great deal of discussion that day about "life is a moving target."

It wasn't long after my trip to Joe's house that I visited my first Consumer Electronics Show in Las Vegas, around 1978, and observed firsthand the wealth of technological change that was about to hit us. I was among the excited crowd looking over an exhibit of a breakthrough Apple personal computer. I was becoming hooked on studying change as a key to the future, particularly changes in technologies and global demographics, and the effect of these changes on the world of business.

A Different Slant on Change: "Nothing is New"

About the same time, I had the opportunity to learn a different take on change, as I traveled on several occasions with a worldly man named Robert Slaton. Slaton had recently retired as the head buyer for the world's largest jewelry firm, which at the time was Zales, owner of nearly 1,000 stores ranging from mid-range Zales stores to high-end Bailey, Banks & Biddle locations that sold luxury jewelry and watches. Zales dominated the retail jewelry sector then like Wal-Mart dominates the discount store sector today. My friend Slaton probably knew more about the manufacturing and marketing of jewelry and gift items than anyone on the planet. Bob had visited more countries on buying trips than anyone I have ever met, and he was a wealth of entertaining information about hotels, restaurants, importing, exporting and the global supply chain. He knew how international trade worked, how manufacturers in Italy used simple machinery to crank out gold chains by the yard and how diamond dealers in Israel, Bangkok and Brussels made fortunes buying low and selling high. I was doing initial work on a business startup that would sell jewelry and gifts via a high-end catalog. Slaton took me to New York City's jewelry district, guided me through jewelry manufacturing plants in Massachusetts and Rhode Island that had been owned by the same families for generations, and showed me how to import merchandise in quantity from Germany and Japan. He introduced me to the top people at Zales, and Donald Zale, the firm's chief executive, took me around on a few of his regular, casual Saturday tours of malls and company-owned stores. Don would occasionally pick me up at my house in his Mercedes convertible and whiz me around Dallas-Fort Worth, showing me which store locations within the malls were "power spots," and letting me watch and learn on his store visits. He ran a company that knew how to adapt to changes in both consumer desires and societal needs. For example, Zale Corporation was a pioneer in offering on-site employee child care, and it had been exceptionally successful at making the change from running stores in old-fashioned shopping centers to becoming the dominant jewelers in modern malls. I learned a great deal about merchandising and retailing in a compressed period of time from these men. They also gave me an early look at the power of globalization, as so much of their sourcing was done overseas.

Slaton may have reached retirement age, but he was not lacking for energy, ideas or enthusiasm. Bob had a different slant on the concept of change. One of his favorite sayings was "nothing is new." His point was that a lot of today's "new" ideas or products are actually reworks or advancements of ideas that have been around for decades or even centuries. Today, in other words, Bob might suggest that the Internet is just an advancement of the telephone and its predecessor the telegraph. To him, a modern-day mall was simply an advancement of ancient bazaars, aided by air conditioning and electric lights, and modern catalogs mailed to your home were an evolution from the door-to-door peddlers of old. One of Bob's strengths as a merchandiser was to take an old concept like a locket or a charm bracelet or pocket watch, create what he called a "story" around it and remake it as something modern and popular. "Nothing is new, but everything changes constantly," he would remind me as he showed me how some popular item had an ancient lineage. In the mid-1970s, before the concept became commonplace, Bob tried to get me interested in selling water in single-serving bottles. He brought me more than 20 different brands of local water, in small bottles from Europe, and tried to get me to pick one or two to import and sell to grocery stores. "There's a growing market for little, luxury personal indulgences like specialty waters," he said, (he would have understood the power of the Starbucks concept in a heartbeat) "and people are growing more concerned about the quality of the water they drink." Of course, he was right as usual, but being a seller of bottled water wasn't in my plans. Too bad for me, since it became a multi-billion dollar market in America soon enough. "Find a way to make an old concept better," he'd say, "sell people on change, and you can make a fortune."

In the end, I realized that Sugarman and Slaton were both right. The fact that change is inevitable is a vital thing to remember. It is also worthwhile to remember that many old ideas that were once discredited eventually return to enjoy new success, although their reappearance may be in highly altered form. For example, the virtues of savings and thrift are now back in fashion after many years of profligate spending. The status quo of 15 and 20 years ago has already been shattered by vast changes in many key areas: the amazingly rapid, global adoption of the cellphone; the digitization of entertainment; and intense global competition from Brazil, China, India and other emerging nations, to name but a few of the more interesting developments. Immense changes will be fostered over the near future by trends discussed in this book: always-on communications, the growing middle class, advancing global trade, convergence of key technologies, advances in sectors ranging from nanotechnology to biotechnology to energy to health sciences to education, and significant changes in demographics. These developments hold the key to the Next Boom.

Internet Research Tips

Plunkett's Next Boom video for Chapter Nine:
www.plunkettresearch.com/NextBoom/Videos

Society at a Glance 2009: OECD Social Indicators:
www.OECD.org/els/social/indicators/SAG offers a look at the quality of life in the nations that enjoy the world's largest economies

KIPP: The KIPP Knowledge is Power Program for charter schools can be found at www.kipp.org

National Alliance for Charter Schools: www.publiccharters.org provides information, facts and figures about charter schools

Cisco Systems, Inc. papers and videos on the transformative power of technology in education, "Connected Learning Societies"
www.cisco.com/web/strategy/education

Join in the discussion!

- See the Reading Group Guide in the back of the book.
- Go to Facebook, search for The Next Boom.
- Join The Next Boom group on LinkedIn.

— epilogue —

2026

"Sooner or later, we sit down to a banquet of consequences,"
-Robert Louis Stevenson

New Year's Day, 2026

Imagine that you wake up on the morning of New Years Day, 2026. At precisely 8:55, you begin to stir thanks to the sound of the grinder whirring automatically in your coffee maker. Groggily, slowly becoming aware of the fact that today is a holiday, you get out of bed, motivated by the aroma of brewing coffee. Knowing that you would be out late the previous evening, you set your UniComm to remain dormant until 9:00, far later than your usual wake up hour. Your nearly new house is a technical marvel, typical of other homes built in 2024. The window shades in your bedroom and living spaces opened automatically at 9:00 when the rest of the house began to come to life per your instructions. High efficiency OLED (organic light emitting diode) lighting turned on the second your feet hit the bedroom floor, but dimmed and turned off when you left for the kitchen, where lighting began to glow upon your arrival.

Just like your faithful Labrador Retriever, now wagging her tail, a UniComm screen sprang to life when you entered the kitchen. Communicating with your household UniServer (brimming with nanotech-based chips, and running on amazingly low electricity needs), your UniComm has arranged the news sources that you value most on its bright, flat screen. Yes, there are several messages flashing across the bottom of the monitor, and you elect to have a few voice mails converted to text to read at your convenience later on your UniCommMobi. You check a few "Happy New Year" messages left by friends, some of whom were clearly partying with enthusiasm when they left video calls which you watch with a grin while you pour the coffee.

Sitting down to peruse the news, you begin, as you do every morning, with *AllWorld Today*, the brilliantly successful e-newspaper, Gannett's global version of *USA Today*. You check the headlines and weather for Cuba, since you are leaving later that afternoon to spend a week at your newly acquired beachfront condo (in the exclusive, unplugged development known as "E-Free City") on the coast near Havana. Ahhh, just the thought of it relaxes you. Ever since Fidel finally expired and his brother Raul mysteriously disappeared during a period of violent protests over food rationing, the country has been run by a general of the army who modeled the now-emerging Cuba on the business-friendly platform adopted many years earlier by the communist-governed nation of Vietnam. Cuban entrepreneurship is soaring, construction is booming, and foreign investment is pouring in, particularly from America and Canada. New, deep-water oil fields several miles off the Cuban coastline in the Gulf of Mexico are filling government coffers and creating jobs. The Havana Stock Exchange opened last year. Now, both Cuba and Vietnam are exchanging communism for commercialism, while increasing the powers of local officials, and the Cuban general has promised free elections in 2027.

Next, you scan the global business headlines. The usual New Years Day recaps have been posted by journalists striving to summarize conditions as of year-end 2025. You see that Ghana, Kenya, Mexico, Poland and Turkey likely ended the year as the world's fastest-growing economies. China's GDP growth is down from its soaring rates of the early 2000s, but enviable nonetheless. Wages and median household income in China have grown substantially, one writer reminds you. India, Malaysia, Indonesia and Brazil are likewise maturing steadily, with booming middle classes. A continuing stream of strikes and worker demands for higher pay have eroded the labor cost advantages that these nations enjoyed for many years, while greater household income and a new generation of consumers are supporting vibrant retail markets.

On the blogs side of your screen, you are sad to see that your favorite business observer, Simon X. Lee, is retiring his commentary with today's posting. He began writing his blog on January 1, 2008 in the midst of the Great Recession. (Of course, you know that "Simon X. Lee" is an assumed name, and you tend to believe the rumors that he was formerly a top derivatives trader at the defunct investment house, Bear Stearns.) Simon will retreat to his villa near Chiang Mai in the hills of Thailand, where he will put the finishing touches on his new book (a brief history of entrepreneurship in the U.S., in honor of America's 250th Birthday, to be published in electronic and 3D videobook form in a few months on July 4, 2026).

Today, however, Simon takes advantage of this last column to make extensive comments on the changes he has observed in America, including America's evolving relationship with the rest of the world over these 17 years. He reminds his readers what inspired him to start his column: 2008 through 2010 added up to one of the most confused and disturbing periods in economic history, he recounts. European governments were trying to become more like the America of old—slashing government spending and entitlements while looking for ways to foster business startups and employment. Meanwhile, America was becoming more like a European welfare state—ballooning the federal government's spending until it was 25% of GDP, incurring vast amounts of government debt and displaying unbridled anti-business sentiment, adding up to an environment of little hiring and a disparaging lack of confidence among corporate executives, business owners and investors. Angered, and outspoken by nature, Simon became a daily, and often controversial, commentator online.

After a rocky start when the Great Recession finally bottomed out, he recalls, the American economy regained momentum over several years, enjoying an intense technology- and export-driven boom, while the rapidly expanding American population soaked up excess housing inventory that had been holding the economy back since 2008. The path that finally led to the Next Boom was neither easy nor always steady, he admits. While American businesses were adjusting to consumers who spent less and saved more, a painful financial realignment in Europe was followed by a popping bubble in Asia. Soon enough, however, investors became much more tolerant of risk and excited about global prospects, while a steady stream of advances in nanotech, biotech and communications technologies poured from laboratories onto world markets, and consumers in Asia and Latin America created demand for a wide range of goods and services, fueling revenues for multinational companies.

Today, Simon is particularly pleased to see that manufacturing is, to a growing extent, "reshoring," or returning to the U.S., partly because of the high transportation costs of moving goods from overseas factories back to American markets. However, there are beginning to be many other positive factors for American manufacturing, he points out. At the same time that labor costs rose dramatically in emerging nations, a new generation of factory automation slashed costs and boosted productivity at American plants, and just-in-time inventory was expertly adopted by many types of businesses, encouraging local production. Meanwhile, regional centers of research, trade and manufacturing evolved in cities like College Station, Texas and Nashville, Tennessee that feature access to thriving universities with exciting research labs. Leaders in these cities learned to build synergies between their universities and their business communities, based on tax incentives, a coordination between community colleges and manufacturers that ensured the availability of workers with needed skills, and low interest rates on loans from the Made in America Bank, recently established by the federal government. Starting in 2014, new free trade agreements between the U.S. and key Latin American and Asian nations further boosted American manufacturing and exports.

As an Asiaphile who has long felt that China and America would increasingly grow to rely on each other, Simon states his opinion that China's relationship with the U.S. has developed to a point of mutual, but guarded, respect, as Chinese investment in American factories, along with U.S. investment in Chinese retailing and consumer goods firms, grew to immense proportions. Chinese brands such as Haier are widely accepted in North America now, Simon points out, in the same way that Japanese brands such as Sony and Honda grew popular beginning in the 1960s. Today, a significant amount of goods bearing Chinese brands are manufactured in North America, particularly in Mexico, former American "Rust Belt" states like Ohio (thanks to huge tax incentives), and the Southeastern United States. Meanwhile, American food, personal care products, clothing and entertainment brands are earning soaring profits in China, as U.S. firms have learned the secrets to becoming fashion leaders, trend setters and makers of consumer products localized just enough to be perfect for Chinese tastes.

Simon Lee goes on to comment on the wrenching changes he has observed in U.S. government. An important catalyst, he says, was the firebrand female senator from Oklahoma who, a few years after the Great Recession ended, railed against what she called, "the nanny-statization of America," warning that, "excessive anti-business legislation, multiple regulatory burdens, unpayable entitlement commitments and an absurdly complicated and costly income tax system could, if we failed to enact reforms, impede the future of all Americans," in the same way they had created difficult economic conditions and a flight of employers from numerous locales where state governments had been hyperactive spenders, borrowers and regulators.

Taxpayer angst had been growing for years, and was brought to the boiling point when a major news outlet ran a nightly feature for a full week, highlighting the history of government debt and spending once the federal debt had reached $100,000 per working-age member of the population. Shortly thereafter, members of Congress from both sides of the aisle were jolted by protests and demonstrations by millions of taxpayers nationwide when the federal Office of Management and Budget glumly announced that, by 2021, yearly interest costs on federal indebtedness would top $1 trillion for the first time. In other words, Simon explains, deficits and borrowing had been so high that interest alone cost as much as the entire federal budget of 1990. Cost cutting, tax reform and federal streamlining of unprecedented proportions ensued, fueling tremendous leaps in stock market indexes.

Energy conservation and production have evolved dramatically in America during recent years, Simon reminds you. A handful of new nuclear generation plants finally received funding and now are either in operation or under construction, and the Yucca Mountain, Nevada nuclear waste storage facility finally opened last year after fees offered to the State of Nevada were too rich to resist and state leaders ended their blocking tactics. Meanwhile, five companies have filed applications to build the first small, local nuclear plants, each based on slightly different technologies that will create ultra-safe, relatively inexpensive nuclear facilities about the size of a railcar that will provide power to about 20,000 homes. Production from bounteous shale gas fields continues to hold natural gas prices at reasonable levels, while a significant percentage of engines in the nation's largest truck fleets now run on natural gas. Plug-in electric cars now account for a significant portion of all autos purchased in the U.S., with the leading brand being China's BYD, lately manufactured at high capacity at this firm's plant in Tennessee. While the long-sought goal of electric cars with 500-mile range has not yet been reached, Simon says, the latest models do offer nearly 400-mile range, thanks to lithium-air battery technology pioneered by nano-scale researchers at IBM. Next, Simon fumes a bit about energy policy. America's ethanol technology has finally evolved from the preposterous use of a food crop, corn, as a base for ethanol, to the use of quick-growing algae and grasses. Nonetheless, ethanol, like wind power, continues to need an immense federal subsidy, and the continuing progress of electric and natural-gas transport will eventually marginalize ethanol, Simon forecasts. He asks, "What could we have done with that $200 billion we wasted through 2025 on ethanol subsidies? Replace the entire long-distance electric grid with high-efficiency, smart systems? Build a dozen new nuclear generating plants plus an advanced infrastructure of nationwide outlets for recharging electric car batteries?"

In his next-to-last paragraph, Simon takes one final opportunity to boost his personal civic passion. He pours out his praise and thanks to America's charter school pioneers. Growing steadily from a mere 1.5 million students in 2010, he reminds his readers, America's charter schools, both online and brick and mortar, provided an education to 15 million students last year, or about one-quarter of all primary and secondary school students. Taxpayers had reached the tipping point on public education, according to Simon's interpretation of events, unwilling to continue to fund unproductive, ineffective traditional schools where obstructionist teachers' unions and archaic regulations encouraged decline rather than progress. Once the significant benefits of charter schools became common knowledge, state legislators buckled to taxpayer demands for true education reform. Ticking off the results, he reminds us once again that standardized test scores are up dramatically, the drop-out ratio plummeted, personal safety at schools has improved and the percentage of students enrolling in college or technical school after graduation from charter schools hit a new record in 2025. Much of this success came at a low cost per student, he says, thanks to fewer layers of bureaucracy and big advances in online learning and in-classroom technology. Meanwhile, surveys show that teachers are more satisfied with their careers, and the best-performing teachers have earned substantially higher pay.

Simon can't help reminding you that the world's population passed the 7 billion mark in 2012, and will very soon pass 8 billion. Fortunately, America has retained its population growth advantage, and young, legal immigrants and well-documented guest workers continue to bring much needed balance to the aging portion of the American population thanks to the Immigration Reform Act of 2016. Thanks to this new law, foreign students who receive graduate degrees at American universities can be put on a high-speed path to green cards and permanent residency.

Writing on, Simon points out an interesting change brought by globalization: with the recent retirement of Warren Buffett, the investment genius and sometimes world's richest person referred to for decades as the "Oracle of Omaha," the financial guru's mantle has passed to the "Maestro of Mumbai," a 48-year old in India who amassed several billion dollars in a short period of time by running a unique fund focused on investments in India's infrastructure and cost-saving medical technologies. The Maestro of Mumbai seems quite happy with his fame—publishing anxiously awaited white papers on the "state of the world" on the morning of each year's Diwali (the Hindu festival of Lights), and popping off the occasional sound bite or quotable quip for major news outlets. The Maestro has attained near godlike status to the hordes of newly minted MBAs and aspiring entrepreneurs in emerging nations. Adding to India's financial imprimatur, Simon notes, is the fact that the world's two richest people are no longer Americans, but industrialists in India who each own global businesses ranging from consulting to steel to automobiles, wireless telecommunications, hotels, biotechnology and discount retailing under their corporate mantles. Wouldn't it be interesting, Simon muses, if the Indian billionaires, when the time comes, seek to add their fortunes to the Gates and Buffett billions managed by the Bill & Melinda Gates Foundation? Already the world's wealthiest and best-run charitable organization (thanks to the funds and management acumen of its donors), the foundation's gifts and advice have spurred the near-eradication of many of the world's worst diseases, while dramatically boosting the effectiveness of public education worldwide and distributing easy-to-implement technologies that made clean drinking water available for the first time in many of the world's remotest villages.

Reminded of "water," you murmur "environmental news" in the direction of the UniComm, which responds instantly with appropriate news sources. Under today's lead headline, "Experts Disagree over the Cause of the Average Global Temperature Drop in 2025," you check yesterday's closing price of water futures on the Shanghai exchange. It was profound good fortune, you remember, that led you to focus your graduate studies in statistics on historical water use and production, which eventually led you to become a leading expert on current and future water needs. Water would be one of the world's big challenges, you realized as you were defending your dissertation in 2015. Working bleary-eyed 80-hour weeks and forsaking any sort of social life, you created a model that plots population growth and water usage on a nation-by-nation basis, with unique overlays of variables that include input for weather patterns such as *el Niño*, expected changes in agricultural water usage and the effect of emerging water conservation and distillation technologies. While you have been running your hydrostatistics consulting company and supervising your six employees, you've made enough money by investing in water technology firms to pay cash for that Cuban condo and the ranch outside Spokane, Washington where you live and work. You know that water conservation, desalination and reverse osmosis technologies are finally reaching a level where they can be a cost-effective help with the world's water needs. However, the true turning point, you first forecasted several years ago, will soon come when genetically engineered crops and precision agriculture gain enough critical mass to dramatically cut the amount of water used in agriculture—long the world's water hog. Meanwhile, water remains a challenge.

Your news screen is flashing a "special report," led by a photo of an Indian physician and the headline "World's Richest Doctor Opens 100th Hospital." Intrigued by this famous medical entrepreneur, you scan the story and photos regarding his latest hospital in Monterrey, one of 12 he now operates in Mexico, where his revolutionary surgical and diagnostic centers are treating locals as well as medical tourists from the U.S., Canada and South America at modest cost with spectacular outcomes. The story reminds you that your own health insurer will pay your travel expenses and waive your deductible if you have that back surgery done at one of these Mexican hospitals. Of course, you muse, your insurer will save $17,000 over U.S. costs.

"Power off," you mumble at your UniComm. Time to walk the dog and pack for that flight to Cuba.

What Could Go Wrong and Final Optimistic Thoughts

The bit of fiction above is a scenario, not a prediction, but many of the conditions presented in this view of New Year's Day, 2026 could reasonably occur. There are no guarantees about the occurrence, timing or intensity of the Next Boom. Having said that, I firmly believe that the initial groundwork for tremendous economic growth was already being laid by 2010, and I have attempted to document and describe many of the most important trends and technological advancements that will boost the world over the near future, enabling global standards of living, wealth and income to soar. I also believe that the typical consumer, executive or investor consistently underestimates the long term effects of new and evolving technologies. Further enhancements to cutting-edge technologies, in particular wireless communications, nanotech and biotech, as discussed in brief form in this book, are poised to provide an immense boost to the global economy. Converging technologies will also help to alleviate environmental challenges in a cost-effective manner. Sustainability will become more of a reality, and green business practices will become more commonplace, when the costs of green technologies decline thanks to volume production.

I also believe that the Great Recession created a sea change in consumer and voter attitudes. A rapidly rising number of citizens in economically mature nations, particularly America, Japan and the U.K., will demand significant change for the better from their governments, and this will lead to wiser use of government funds; better education systems; stronger support for business investment and technological research; and better nurturing of job-generating startups over the near future. Meanwhile, emerging nations such as Brazil, China and India have barely begun to enjoy the economic boost they are going to see from their soaring business and industrial bases and rising middle classes.

Yes, a lot of things could go wrong, from counterproductive government practices, widespread war, disease or natural disaster to a financial crisis that gets out of control. In addition, as with the Great Boom of 1982-2007, the Next Boom is likely to suffer from interruptions, wrong turns, periods of crisis, ups, downs, disasters and hesitations. A double-dip extension of the Great Recession is at least remotely possible as I write this in mid-2010. European and Japanese governments are facing the need for painful reductions in expenditures and debt. Asian economies have been growing so quickly that they will likely go through temporary bubble-popping adjustments. In America, federal, state and local governments face daunting problems in terms of debt, entitlement obligations, inefficiencies and ineffective public education.

By mid-2010, the first leg of the Next Boom appeared to be well underway in many parts of the world. For America to participate, a trend of substantial growth in new jobs must take hold. For that to happen, businesses must have sufficient confidence in the future to hire and invest. For that to happen, Washington will have to make an abrupt turn away from the blinding fog of uncertainty that enveloped the American business sector in 2010, including: uncertainty over the effects of massive new legislation on health care; uncertainty over the effects of an equally massive new bill covering the entire scope of the financial services industry; uncertainty over income tax rates; and uncertainty about proposed sweeping regulation of the energy industry. This environment has made it exceptionally hard to hire and make bets on the future. Nonetheless, I'll stick with optimism, as I believe a dramatic turn is on the horizon, and I trust in the potential positive effects of entrepreneurship, innovation and hard work over the near future.

The Earth will soon be home to 8 billion people and a global economy in the $100 trillion range, creating intense competition for resources and markets, along with complexities, challenges and opportunities of unimaginable proportions. There is no reason to expect the Next Boom to have an equally beneficial effect on all people or all nations. Over the long run, the countries that benefit the most will be those with government policies and societal conditions that support efficient infrastructure, research and development, modest tax rates, rule-of-law, reasonable access to lending and credit, and well trained workforces, along with relative efficiency and honesty in public administration. Nations that participate fully in the Next Boom will be those that have leaders who proactively prepare for the changes that are afoot—nations that foster conditions that are conducive to the growth of business, employment and investment. In many cases, this will require fresh thinking and painful adjustments. To paraphrase a well-known axiom that relies on "lunch" as a metaphor for economic gain: There's no such thing as a free Boom.

— Reading Group Guide —

What You Absolutely, Positively Have to Know
About the World Between Now and 2025

by: Jack W. Plunkett

Reading Group Guide

Jack W. Plunkett's *The Next Boom* aims to help readers understand and benefit from the coming era of global growth and change. One way to further explore these topics is through discussion with fellow readers (such as friends, classmates, co-workers or business partners). We hope that the following discussion questions will serve to stimulate conversation, encourage readers to share and further develop their ideas and enrich the overall reading experience.

Chapter One

1. Chapter One discusses the large expected growth in the U.S. population in coming decades, as well as the dramatic shifts in demographics that will accompany that growth. How do you think Americans can best prepare themselves for these changes? What sorts of "growing pains" do you think we may face in years ahead?

2. As the American population grows and the demographic make-up of U.S. citizens evolves, what sorts of implications do you foresee for education, especially at the elementary and secondary levels? How will changes in the "average American" create new kinds of students? In what ways might our present educational system be especially taxed by the increase in student numbers and by the influx of new students from different backgrounds?

3. Birth rates tend to decrease as a country becomes more educated, industrialized and prosperous. What is your current image of the "average American family," and how do you think this prototypical family might change in 20, 50, or 100 years? How do these changes affect our understanding of what a family is, and how might they affect our cultural identity over time?

4. Advances in medicine are enabling people to live longer, and the "senior citizen" percentage of the population will continue to grow. How can countries benefit from their growing senior populations, and

how can the nations of the world help to ensure that seniors continue to enjoy quality of life as their life spans increase?

Chapter Two

5. Chapter Two discusses changing consumer habits and the beginning of a "Post-Excessive Consumption Era." Do you agree that the increases in savings and the decreases in debt will be long-term trends? Have we "learned our lesson" or will consumers continue to struggle with excessive spending and debt when the economic environment improves?

6. In what ways has the Great Recession affected how you think about money? In what ways has it affected how you think about work? Consider some of the concrete effects on your own family, your personal finances or your career. Have there been benefits as well as challenges?

7. In what ways might the recession have an impact on today's young people? In particular, consider those finishing high school or college in 2009 or 2010 and entering the adult workforce for the first time.

8. Consider some of the challenges and opportunities presented by the emerging LOHAS segment of the consumer population. How can companies respond to this trend in a way that balances concern for environmental and social issues with the practical need to generate profits and keep costs under control?

Chapter Three

9. Consider the rise of firms like Evalueserve and the concepts of Knowledge Process Outsourcing and Business Process Outsourcing. How do you feel about people all over the planet competing for the same jobs? What might some of the benefits of this trend be, and what are some of the downsides? What advice might you give to your children or to a young person in school regarding the future of work in a globalized economy?

10. As the Next Boom gains momentum and the world becomes an even smaller place (more interconnected and more interdependent), how can the needs of the global economy be balanced against the rights of individual people to their individual beliefs? Even if extremism—religious, political or otherwise—can't be eliminated, how might it be mitigated, or its negative consequences reduced? What kinds of solutions and approaches might corporations consider? What kinds of approaches might have to begin with individuals?

11. Which do you use more these days, a mobile phone or a landline phone? If you aren't among those who have already given up their landline phones entirely, can you imagine yourself doing so any time in the near future?

Chapter Four

12. In the global business environment of the Next Boom and beyond, what will be the role of nationalism and the belief in the superiority of one's own country? Will the barriers between nations become less important as everything becomes more connected? What might be the proper balance between competition and cooperation among the world's nations?

13. What does "middle class" mean to you, and how is it different from being poor or rich? Also, how much is it about actual income, and to what degree is it a state of mind? How does a person determine when enough is enough?

14. In discussing the need for improvements to education in the U.S., the author states that the goal should be to produce more graduates with "marketable job skills in tune with the realities of the world economy." To what extent do you believe this requires a four-year college education, and to what extent might some of America's educational shortcomings be addressed through alternate methods such as two-year colleges, vocational schools and a reinvigoration of the apprenticeship model? Also, why do you think so many American students aren't living up to their educational potential? Why do so many young people drop out of high school and college in the U.S., and how might this issue be most effectively addressed?

Chapter Five

15. In reading Chapter Five, were you surprised to learn how much American agriculture involves genetically modified seeds and crops? How comfortable are you with this trend, and why do you think people in different parts of the world react so differently on this topic?

16. Consider the idea of "needs-based innovation," such as that utilized at the Indian telecom firm Bharti Airtel. Has the recession forced you to apply this principle in areas of your own work, or in your personal life? How do you think entrepreneurs, businesspeople, policymakers and individuals might learn to develop their creative adaptation and problem-solving skills so that they can remain inspired to grow and innovate even in difficult times?

17. One of the themes of this chapter, as well as the book as a whole, is that dedicated entrepreneurship and continued innovation can lead to developments that will make the world a better, healthier and more prosperous place. In a broad sense, the author is choosing to take the stance of an optimist, and warns his readers not to allow present setbacks or fears about the future to discourage them. As you read, do you feel any of your beliefs (about business, the world or life in general) being challenged? Do you find yourself agreeing or disagreeing with the idea that dedicated innovation can continue to make the world a better place? How optimistic are you at this point?

Chapter Six

18. As technological advances continue to allow us to be more productive and to complete many tasks more efficiently, how might we ensure that such advances also improve the *quality* of our lives? As an example, the development of mobile smart phones that connect us 24/7 to our e-mail and to the Internet can be an amazing boon to productivity, but it also means that our work can now follow us practically everywhere, all the time. How might people safeguard themselves against becoming virtual slaves to the technologies that were supposed to bring greater freedom and progress to their lives?

19. When you read about the goals of MIT's Project Oxygen (including sensors embedded in the walls of your home and office, computers that recognize you when you enter a room, and cloud-networked systems filled with information, what is your reaction? Do you find the idea thrilling, or does it bring to mind images from George Orwell's *Nineteen Eighty-Four*, with its representation of a society under constant surveillance?

20. Consider Ray Kurzweil's assertion that technological advancement will continue at an exponential rate throughout the next century, resulting in developments that can hardly be comprehended (or even imagined) today. What do you think of this idea? How do you imagine the world a hundred years from now?

Chapter Seven

21. Did this chapter challenge your assumptions about sustainable growth and future energy needs? After reading, do you feel more hopeful about the possibility that continued innovation and new technologies will be able to converge with human ingenuity to provide adequate and affordable power to a growing world population?

22. Chapter Seven discusses the work of George P. Mitchell, whose company carried on with shale gas exploration efforts despite naysayers and incredible engineering obstacles. Despite the persistence of "the human capacity for pessimism," innovators continue to forge ahead and chart new paths toward the future. How might you (or your company) develop a stronger ability to see opportunity where others primarily see problems and difficulties?

23. The blowout of the Deepwater Horizon drilling rig in 2010 (and the oil spill that followed) brought into sharp focus some of the potential dangers and complications involved in energy exploration and production. Who do you think should be responsible for regulating these efforts? How involved should the federal government be, and in what capacity? How can the demands of safety and caution be suitably weighed against the necessary risks?

Chapter Eight

24. When some health care costs are paid for by the government or through employers, how much right does the government or the employer have to attempt to control an individual's behavior in areas that might affect those costs? Should an employer be able to penalize an employee for tobacco use if the employee is already paying higher insurance premiums because of his choice to smoke? Should the federal government be able to put an overweight person on a diet if some or all of that person's health care costs are being covered through government sources?

Chapter Nine

25. The developing U.S. charter school movement is discussed in this chapter. How far can this model be taken as a solution to America's widespread educational difficulties? What role do you think charter schools might be able to play in overall education reform?

26. Consider the discussion of generations and generational differences. What does your age and generation say about you? How do you differ in outlook from someone 20 years younger or 20 years older? Discuss some of the implications, for businesses and for society, of the evolving mindsets of today's younger generations. What do you think our children will do with the legacies handed on to them?

Notes

Introduction

[1] Byron W. King, "One More Oil Boom," *Whiskey and Gunpowder*, September 9, 2005, http://whiskeyandgunpowder.com/oil-booms-one-more-oil-boom (accessed February 22, 2010).

[2] U.S. Census Bureau, Population Division, "US & World Population Clock," *See* http://www.census.gov/main/www/popclock.html (accessed June 24, 2010).

[3] Investopedia, "What Is the 'Rule of 72?'" *See* http://www.investopedia.com/ask/answers/04/040104.asp (accessed June 23, 2010).

[4] National Bureau of Economic Research, Business Cycle Dating Committee, "Determination of the December 2007 Peak in Economic Activity," December 11, 2008, *See* http://www.nber.org/cycles/dec2008.html (accessed June 22, 2010).

Chapter One

[1] All United States population figures are from U.S. Bureau of the Census reports unless otherwise stated. Official Census figures used in this book may not yet fully integrate the effects of the recent recession. In addition, the final results of the 2010 Census may cause significant changes in official projections.

[2] Texas Department of State Health Services, Center for Health Statistics, "Projected Texas Population by Area, 2010," July 2, 2009, *See* http://www.dshs.state.tx.us/chs/popdat/ST2010.shtm (accessed June 24, 2010).

[3] Robert E. Lang, Mariela Alfonzo and Casey Dawkins, "American Demographics - Circa 2109," *Planning*, May 2009, 10-15.

[4] Population Division of the Department of Economic and Social Affairs of the United Nations Secretariat, *World Population Prospects: The 2008 Revision Population Database,* http://esa.un.org/UNPP (accessed July 15, 2010).

[5] *State of the Union's Finances* (Peter G. Peterson Foundation, March 2009).

[6] Julian L. Simon, *The Ultimate Resource 2* (Princeton: Princeton UP, 1996), 326.

[7] Conor Dougherty and Miriam Jordan, "Recession Hits Immigrants Hard," *Wall Street Journal*, September 23, 2009, A10.

[8] "Suing Arizona," *Wall Street Journal*, July 7, 2010, A16.

[9] "The decennial census: The knock on the door," *The Economist*, June 13, 2009, 34.

[10] Jonathan Wald, "U.N.: Population to Top 9 Billion by 2050," *CNN.com*, February 25, 2005, http://edition.cnn.com/2005/US/02/24/un.population (accessed June 25, 2010).

[11] George Friedman, *The Next 100 Years: A Forecast for the 21st Century* (New York: Doubleday, 2009), 51.

[12] Population Division of the United Nations.

[13] Lee Kuan Yew, "Changes in the Wind," *Forbes*, October 19, 2009, 31.

[14] Simon Shuster, "Russia's Medvedev Launches a New War on Drinking," *TIME*, August 23, 2009, http://www.time.com/time/world/article/0,8599,1917974,00.html (accessed July 1, 2010).

[15] Richard Dobbs, James Manyika, Charles Roxburgh and Susan Lund, *Lions on the Move: The Progress and Potential of African Economies* (McKinsey Global Institute, June 2010), vii.

[16] Randall D. Schnepf, Erik Dohlman and Christine Bolling, "Agriculture in Brazil and Argentina: Developments and Prospects for Major Field Crops," Economic Research Service/USDA, Report no. WRS-01-3, November 2001.

Chapter Two

[1] Visa Investor Relations, "Total U.S. Visa Debit Volume Surpasses Credit for First Time," Visa, Inc. press release, May 4, 2009, http://corporate.visa.com/media-center/press-releases/press950.jsp (accessed July 29, 2009).

[2] Aparajita Saha-Bubna, "AmEx Net Drops 48% As Charge-Offs Increase," *Wall Street Journal*, July 24, 2009, http://online.wsj.com/article/SB124838009658476871.html (accessed July 16, 2010).

[3] American Express, "How the Charge Card Can Help Remove Debt From Your Wallet," advertisement, *New York Times*, September 4, 2009, A5.

[4] "A special report on America's economy: Time to rebalance," *The*

Economist, April 3, 2010, 3-7.

[5] Reuven Glick and Kevin J. Lansing, "U.S. Household Deleveraging and Future Consumption Growth," *FRBSF Economic Letter* no. 2009.16 (2009), Federal Reserve Bank of San Francisco, May 15, 2009, 1-3.

[6] Robert Reich, "The Jobs Picture Still Looks Bleak," *Wall Street Journal,* April 12, 2010, http://online.wsj.com/article/SB10001424052702304222504575173780671015468.html (accessed July 16, 2010).

[7] American Express Investor Relations, "Majority of Consumers Report No Increase in Debt Over Past Six Months According to American Express Spending & Savings Tracker," American Express Co. press release, July 13, 2010, http://about.americanexpress.com/news/pr/2010/axp_tracker2.aspx (accessed July 16, 2010).

[8] Peggy Noonan, "Common Sense May Sink ObamaCare," *Wall Street Journal,* July 25, 2009, A13.

[9] David Brooks, "The Great Unwinding," *New York Times,* June 12, 2009, A27.

[10] Eswar S. Prasad, "Rebalancing Growth in Asia," IZA Discussion Paper no. 4298, July 2009, Table 1.

[11] David Wessel, "Asia's Latest Export: Recovery," *Wall Street Journal,* February 24, 2010, http://online.wsj.com/article/SB10001424052748703510204575085280515242598.html (accessed July 16, 2010).

Chapter Three

[1] Population Reference Bureau, "2008 World Population Data Sheet," August 19, 2008, *See* http://www.prb.org/pdf08/08WPDS_Eng.pdf (accessed July 27, 2010).

[2] Lance E. Davis, *American Economic History: The Development of a National Economy* (Homewood, IL: R.D. Irwin, 1961), 283, Table 15-1.

[3] Katrin Bennhold, "Chinese Leader Offers a Glimpse of the Future," *International Herald Tribune,* January 28, 2010, A1.

[4] World Bank Group, *Global Economic Prospects 2007: Managing the Next Wave of Globalization* (Washington, DC: World Bank Publications, December 2006), xiv.

[5] Heather Timmons and Hari Kumar, "India Steadily Increases Its

Lead in Road Fatalities," *New York Times,* June 9, 2010, A4.
[6] Eric Bellman, "India's Outsourcing Firms Lure More Japan Business," *Wall Street Journal,* August 17, 2009, B1.
[7] World Bank Group, xiii.
[8] Christine Zhen-Wei Qiang, "Mobile Telephony: A Transformational Tool for Growth and Development," *Private Sector & Development* 4, Proparco's Magazine, November 2009, 8.
[9] Prashant Malaviya, Arvind Singhal, P.J. Svenkerud and Swati Srivastava, *Telenor in Bangladesh (Case B): Achieving Multiple Bottom Lines at GrameenPhone*, (INSEAD - The European Institute of Business Administration Report, March 2004), 4.
[10] World Trade Organization, "Regional Trade Agreements: Facts and Figures," *See* http://www.wto.org/english/tratop_e/region_e/regfac_e.htm (accessed July 27, 2010)
[11] Timothy Aeppel, "Coming Home: Appliance Maker Drops China to Produce in Texas," *Wall Street Journal,* August 24, 2009, B1-B2.

Chapter Four

[1] Jayson Chi, Osamu Kaneda, Gordon Orr and Brian Salsberg, *Winning the Chinese Consumer: Opportunities for Japanese Companies* (McKinsey & Company, September 2009), 12.
[2] "Selling Foreign Goods in China: Impenetrable," *The Economist,* October 17, 2009.
[3] Christoph Rauwald, "Upscale Car Makers See Sign of Hope," *Wall Street Journal,* November 10, 2009, B2.
[4] Eric D. Beinhocker, Diana Farrell and Adil S. Zainulbhai, "Tracking the Growth of India's Middle Class," *McKinsey Quarterly,* 3 (2007): 52.
[5] "Who's in the Middle?" *The Economist*, February 12, 2009, http://www.economist.com/node/13063338?story_id=13063338 (accessed on April 3, 2009); Branko Milanovic and Schlomo Yitzhaki, "Decomposing World Income Distribution: Does The World Have A Middle Class?" *Review of Income and Wealth* 48, no. 2 (June 2002).
[6] World Bank Group, *Global Economic Prospects 2007: Managing the Next Wave of Globalization* (Washington, DC: World Bank Publications, December 2006), xvi.
[7] Ibid., xiii.
[8] Chi, et al., 12.

[9] Mark Schneider, "The Costs of Failure Factories in American Higher Education," *Education Outlook*, 6 (October 2008).

[10] U.S. Census Bureau, *School Enrollment in the United States: 2006*, Table A-5a. *See* http://www.census.gov/population/socdemo/school/TableA-5a.xls (accessed July 27, 2010).

[11] Stephen S. Roach, *The Next Asia: Opportunities and Challenges for a New Globalization* (Hoboken: John Wiley & Sons, 2009), 156.

[12] World Trade Organization, *World Trade Organization Annual Report 2007* (Geneva, Switzerland: World Trade Organization, 2007).

Chapter Five

Note: *Plunkett's Engineering & Research Industry Almanac 2011* served as a primary source for much of this chapter. Interested readers who seek more in-depth analyses for topics touched on here may find it useful.

[1] Robynne Boyd, "Genetically Modified Hawaii," *Scientific American*, December 8, 2008, http://www.scientificamerican.com/article.cfm?id=genetically-modified-hawaii (accessed July 20, 2010).

[2] Gregg Easterbrook, "The Man Who Defused the 'Population Bomb,'" *Wall Street Journal*, September 16, 2009, A27.

[3] "2050: A third more mouths to feed," Food and Agriculture Organization of the United Nations press release, September 23, 2009, http://www.fao.org/news/story/0/item/35571/icode/en (accessed July 20, 2010).

[4] Venessa Wong, "A Big Idea for Little Farms," *BusinessWeek*, November 23, 2009, 53.

[5] "Feeding the world: If words were food, nobody would go hungry," *The Economist*, November 21, 2009, 62.

[6] "Monsanto: The parable of the sower," *The Economist*, November 21, 2009, 73.

[7] Niu Shuping and Tom Miles, "China gives safety approval to GMO rice," *Reuters*, November 27, 2009, http://www.reuters.com/article/idUSPEK37812 (accessed August 12, 2010).

[8] Shaobing Peng, "Challenges for rice production in China," *Rice Today* 6, no. 4 (October-December 2007): 38.

[9] Norman E. Borlaug, "Farmers Can Feed the World," *Wall Street Journal*, July 31, 2009, http://online.wsj.com/article/SB10001424052970203517304574304562754043656.html (accessed July 30, 2010).

[10] Clive James, "Global Status of Commercialized Biotech/GM Crops: 2009," International Service for the Acquisition of Agri-Biotech Applications (ISAAA), Issue Brief 41-2009, 2009.

[11] "Monsanto: The parable of the sower," 71.

[12] Robert Langreth and Matthew Herper, "The Planet Versus Monsanto," *Forbes*, January 18, 2010, 69.

[13] Ibid., 67.

[14] Andrew Pollack, "In Lean Times, Biotech Grains Are Less Taboo," *New York Times*, April 21, 2008, http://www.nytimes.com/2008/04/21/business/21crop.html (accessed August 12, 2010).

[15] Jack Kaskey, "Monsanto Sets a Soybean Free," *BusinessWeek*, February 8, 2010, 19.

[16] Dan Collins, "Anti-Biotech Protests Grow Violent," *CBS News*, October 15, 2003, http://www.cbsnews.com/stories/2003/08/06/tech/main566892.shtml (accessed March 2, 2010).

[17] Spencer E. Ante, "Fertile Ground for Startups," *BusinessWeek*, November 23, 2009, 48.

[18] Pete Engardio, "Innovation Goes Downtown," *BusinessWeek*, November 30, 2009, 52.

[19] "2010 Global R&D Funding Forecast," *R&D Magazine*, December 2009, 3.

[20] Adrian Slywotzky, "How Science Can Create Millions of New Jobs," *BusinessWeek*, September 7, 2009, 37.

[21] Priya Ganapati, "Bell Labs Kills Fundamental Physics Research," *Wired.com*, August 27, 2008, http://www.wired.com/gadgetlab/2008/08/bell-labs-kills (accessed August 12, 2010).

[22] Jeffrey R. Immelt, Vijay Govindarajan and Chris Thimble, "How GE Is Disrupting Itself," *Harvard Business Review* (October 2009): 3.

[23] Jason Pontin, "Q&A: Bill Gates," *Technology Review*, September/October 2010, 32.

Chapter Six

[1] Barnaby J. Feder, "Richard E. Smalley, 62, Dies; Chemistry Nobel Winner," *New York Times*, October 29, 2005, http://www.nytimes.com/2005/10/29/science/29smalley.html (accessed March 12, 2010).

[2] Project on Emerging Nanotechnologies, "Nanotechnology Inventories," *See* http://www.nanotechproject.org/inventories (accessed

August 19, 2010).

[3] Jack W. Plunkett, *Plunkett's Nanotechnology & MEMS Industry Almanac 2010* (Houston: Plunkett Research, Ltd., 2010), 8.

[4] Jack W. Plunkett, *Plunkett's E-Commerce & Internet Business Almanac 2010* (Houston: Plunkett Research, Ltd., 2010), 29.

[5] Philip Shapira and Alan Porter, "Nanotechnology: Will it Drive a new Innovation Economy for the U.S.?" (Presentation, Project on Emerging Nanotechnologies, Washington, DC, March 23, 2009).

[6] David Schrank and Tim Lomax, *2009 Urban Mobility Report* (Texas Transportation Institute at Texas A&M University, July 2009), 5.

[7] "HP Enables Better, Faster Decision Making with Breakthrough Sensing Technology," Hewlett-Packard Development Company, L.P. press release, November 5, 2009, http://www.hp.com/hpinfo/newsroom/press/2009/091105xa.html (accessed March 9, 2010).

[8] *Connecting America: The National Broadband Plan* (Washington, DC: Federal Communications Commission, March 16, 2010). *See* http://www.broadband.gov/download-plan (accessed August 23, 2010).

[9] "Over 1 Billion Global Broadband Subscribers by 2013," InStat, LLC press release, January 20, 2010, http://www.instat.com/press.asp?ID=2706 (accessed August 18, 2010).

[10] MIT Project Oxygen, Computer Science and Artificial Intelligence Laboratory, "MIT Project Oxygen: Overview," *See* http://oxygen.lcs.mit.edu/Overview.html (accessed August 19, 2010).

[11] International Technology Roadmap for Semiconductors, "ITRS 2009 Edition Executive Summary," December 2009, *See* http://www.itrs.net/links/2009ITRS/2009Chapters_2009Tables/2009_ExecSum.pdf.

[12] Ray Kurzweil, "The Law of Accelerating Returns," *KurzweilAI*, March 7, 2001, http://www.kurzweilai.net/the-law-of-accelerating-returns (accessed August 19, 2010).

Chapter Seven

Note: *Plunkett's Energy Industry Almanac 2010* and *Plunkett's Renewable, Alternative & Hydrogen Energy Industry Almanac 2010* served as significant sources for this chapter. Interested readers who wish to learn more about the topics touched on in this chapter may find them beneficial.

[1] Federal Energy Regulatory Commission, *High Natural Gas Prices: The Basics* (U.S. Department of Energy, December 8, 2005), 2. *See* http://www.ferc.gov/legal/staff-reports/high-gas-prices.pdf

[2] Oilnergy.com , "NYMEX Henry-Hub Natural Gas – Daily Price," August 25, 2010, *See* http://www.oilnergy.com/1gnymex.htm (accessed August 26, 2010).

[3] "Mitchell Gift to Endow Academies' Efforts in Sustainability Science," *In Focus Magazine* 2, no. 2 (Summer/Fall 2002). *See* http://www.infocusmagazine.org/2.2/spotlight.html

[4] Mitchell Energy & Development Corporation, "Company History," *See* http://www.fundinguniverse.com/company-histories/Mitchell-Energy-and-Development-Corporation-Company-History.html (accessed August 31, 2010).

[5] Jim Fuquay, "Q&A George Mitchell, Founder of Mitchell Energy: Barnett Shale pioneer is honored," *Fort Worth Star Telegram*, April 2, 2009, C1.

[6] Energy API, "Facts about Shale Gas," *See* http://www.api.org/policy/exploration/hydraulicfracturing/shale_gas.cfm (accessed April 7, 2010).

[7] *Fueling North America's Energy Future: The Unconventional Natural Gas Revolution and the Carbon Agenda (Executive Summary)* (IHS Cambridge Energy Research Associates, 2010), ES-1.

[8] Ibid., ES-4.

[9] Clifford Krauss, "Forget Wind. Pickens Turns Focus to Gas.," *New York Times*, January 14, 2010, B1.

[10] Ben Casselman, "Firms See 'Green' in Natural-Gas Production," *Wall Street Journal*, March 30, 2010, B7.

[11] Kristen Hays, "George Mitchell still pushes energy conservation," *Houston Chronicle*, August 1, 2008, http://www.chron.com/disp/discuss.mpl/business/5920511.html (accessed August 31, 2010).

[12] "Report Provides Strategy for Transition to Sustainability; Actions Required Over Next Two Generations," National Academies press release, November 9, 1999, http://www8.nationalacademies.org/onpinews/newsitem.aspx?RecordID=9690 (accessed August 31, 2010).

[13] U.S. Energy Information Administration, *Annual Energy Review 2009* (U.S. Department of Energy, August 2009), 13, Table 1.5.

[14] Spencer Swartz and Shai Oster, "China Tops U.S. in Energy Use,"

Wall Street Journal, July 18, 2010 (accessed August 26, 2010).

[15] Richard Newell, "Annual Energy Outlook 2010," (Presentation, The Paul H. Nitze School of Advanced International Studies, Washington, DC, December 14, 2009).

[16] Nano Petroleum, Gas and Petro Chemical Industries Conference, "Oil & Gas challenges for the Nanotech community," *See* http://www.sabrycorp.com/conf/npg/09/industry-expert/industry_expert.cfm (accessed April 10, 2010).

[17] American Public Gas Association, "A Brief History of Natural Gas," *See* http://www.apga.org/i4a/pages/index.cfm?pageid=3329 (accessed April 8, 2010).

[18] Walter Williams, "Environmentalists' Wild Predictions," *Capitalism Magazine*, May 7, 2008, http://www.capitalismmagazine.com/science/environment/3432-Environmentalists-Wild-Predictions.html (accessed April 7, 2010).

[19] Robert Bryce, *Power Hungry: The Myths of "Green" Energy and the Real Fuels of the Future* (New York: PublicAffairs, 2010), 96-97.

[20] "B&W unveils modular nuclear power design," *World Nuclear News*, June 10, 2009, http://www.world-nuclear-news.org/NN-BandW_unveils_modular_reactor_design-1006095.html (accessed April 8, 2010).

[21] Jason Pontin, "Q&A: Bill Gross," *Technology Review*, March/April 2010, 10.

[22] Lee Schipper, "Changing the Paradigm of Transportation Efficiency," in *Energy Vision Update 2010: Towards a More Energy Efficient World* (Geneva, Switzerland: World Economic Forum, 2010), 51.

[23] Elisabeth Rosenthal, "Europe, Cutting Biofuel Subsidies, Redirects Aid to Stress Greenest Options," *New York Times*, January 22, 2008, C3.

Chapter Eight

Note: *Plunkett's Health Care Industry Almanac 2011* and *Plunkett's Biotech & Genetics Industry Almanac 2011* served as significant sources for this chapter. Interested readers who wish to learn more about the topics touched on in this chapter may find them beneficial.

[1] Jeffrey E. Williams, "Donner Laboratory: The Birthplace of Nuclear Medicine," *Journal of Nuclear Medicine* 40, no. 1 (January 1999): 16N.

[2] R. R. Wilson, "Radiological Use of Fast Protons," *Radiology* 47, no. 5 (November 1946): 487-91.

[3] "Fermilab Builds Proton Accelerator to Treat Cancer," *Batavia Chronicle*, January 4, 1989, http://history.fnal.gov/Neutron_Therapy/index.html (accessed September 10, 2010).

[4] Richard A. Schaeffer, *Legacy: Daring to Care* (Loma Linda, CA: Legacy Publishing Association, 1990), 39.

[5] Carl J. Rossi, Jr., "Conformal proton beam radiation therapy for prostate cancer: concepts and clinical results," *Community Oncology* 4, no. 4 (April 2007): 236.

[6] Schaeffer, 36.

[7] Rossi, Jr., 236.

[8] Ibid.

[9] Organisation for Economic Co-operation and Development, "OECD Health Data 2010," June 2010, *See* www.oecd.org/health/health data (accessed September 10, 2010).

[10] Centers for Medicare & Medicaid Services, "National Health Expenditure Projections 2009-2019," *See* https://www.cms.gov/NationalHealthExpendData/downloads/proj2009.pdf, 4, Table 1 (accessed September 28, 2010).

[11] Gautam Kumra, Palash Mitra and Chandrika Pasricha, *India Pharma 2015: Unlocking the Potential of the Indian Pharmaceuticals Market* (McKinsey & Company, 2007), 13.

[12] Yuan Ye and Jiang Guocheng, "China unveils health-care reform guidelines," *Xinhua*, April 6, 2009, http://news.xinhuanet.com/english/2009-04/06/content_11138643.htm (accessed April 28, 2010).

[13] Ian Johnson and Jeanne Whalen, "Novartis Mounts Ambitious Push into China," *Wall Street Journal*, November 4, 2009, B3.

[14] Michael Wei, Doug Young and Donny Kwok, "Germany's Bayer sees sharp growth in China medical mkt," *Reuters*, April 27, 2010, http://www.reuters.com/article/idUSTOE63Q03H20100427 (accessed April 28, 2010).

[15] Heather Timmons, "G.E. Chief Sees India Helping Cut Costs of U.S. Health Care," *New York Times,* October 3, 2009, B3.

[16] Nandita Datta, "Made In India For The World," *Outlook*, May 2, 2009, http://business.outlookindia.com/article.aspx?102083 (accessed April 29, 2010).

[17] "A special report on innovation in emerging markets: First break all

the rules," *The Economist*, April 17, 2010, 6.

[18] Ibid., 7.

[19] Geeta Anand, "The Henry Ford of Heart Surgery," *Wall Street Journal*, November 25, 2009, A16.

[20] "A special report on innovation in emerging markets: First break all the rules," 7.

[21] Stephen Cass, "Cheap DNA sequencing will drive a revolution in health care," *Technology Review*, March/April 2010, 63.

[22] Howard Coleman, "Adverse Drug Reactions (ADRs)," Genelex Corporation, July 30, 2010, *See* http://www.healthanddna.com/drug-safety-dna-testing/adverse-drug-reactions.html (accessed May 3, 2010).

[23] "Gene Therapy Appears to Cure Myeloid Blood Disease in Groundbreaking International Study," Cincinnati Children's Hospital Medical Center press release, March 31, 2006, http://www.cincinnatichildrens.org/about/news/release/2006/3-gene-therapy.htm (accessed May 6, 2010).

[24] "Diagnosis Emerges from Complete Sequencing of Patient's Genes," Howard Hughes Medical Institute press release, October 19, 2009, http://www.hhmi.org/news/lifton20091019.html (accessed September 10, 2009).

[25] "Nano-treatment to torpedo cancer," *BBC News*, March 10, 2009, http://news.bbc.co.uk/2/hi/7935592.stm (accessed May 6, 2010).

[26] Ross DeVol, et al., *An Unhealthy America: The Economic Burden of Chronic Disease – Charting a New Course to Save Lives and Increase Productivity and Economic Growth* (Milken Institute, October 2007).

[27] Plunkett Research Estimate.

[28] Plunkett Research Estimate.

[29] Anti-Fraud Resource Center, "The Problem of Health Care Fraud," National Health Care Anti-Fraud Association, *See* http://www.nhcaa.org/eweb/DynamicPage.aspx?webcode=anti_fraud_resource_centr&wpscode=TheProblemOfHCFraud (accessed September 10, 2010).

[30] Plunkett Research Estimate.

Chapter Nine

[1] John C. Goodman and Matt Moore, "Government Spending on the Elderly: Social Security and Medicare," National Center for Policy

Analysis, Report 247, November 30, 2001.

[2] *The 2009 Annual Report of the Board of Trustees of the Federal Old-Age and Survivors Insurance and Federal Disability Insurance Trust Funds,* May 12, 2009, 10, Figure II.D3. *See* http://www.ssa.gov/OACT/TR/2009/tr09.pdf (accessed July 28, 2010).

[3] Anne Shattuck, *Older Americans Working More, Retiring Less,* Carsey Institute, University of New Hampshire, Issue brief 16 (Summer 2010).

[4] Milt Freudenheim, "More Help Wanted: Older Workers Please Apply," *New York Times,* March 23, 2005, A1.

[5] United States National Science Foundation, Division of Science Resources Statistics, *Scientist and Engineers Statistical Data System (SESTAT),* https://sestat.nsf.gov/sestat/sestat.html (accessed July 28, 2010).

[6] George Pratt Schultz and John B. Shoven, *Putting Our House in Order: A Guide to Social Security and Health Care Reform* (New York: W.W. Norton, 2008), 36-37.

[7] Neil Howe and William Strauss, *Millennials Rising: The Next Great Generation* (New York: Vintage, 2000), 4.

[8] Ibid.

[9] National Assessment Governing Board, "2009 Nation's Report Card for Reading Shows Gains in 8th-Grade Scores as 4th-Grade Scores Hold Steady," United States Department of Education, March 24, 2010, http://www.nagb.org/newsroom/release/release-032410.htm (accessed July 28, 2010).

[10] United States Department of Education, Budget Office, *Fiscal Year 2010 Budget Summary and Background Information* (Washington, DC: GPO, May 7, 2009), 4. *See* http://www2.ed.gov/about/overview/budget/budget10/summary/10summary.pdf (accessed July 28, 2010).

[11] Jack W. Plunkett, *Plunkett's Food Industry Almanac 2010* (Houston: Plunkett Research, Ltd., 2010).

[12] *The Economic Impact of the Achievement Gap in America's Schools* (McKinsey & Company, April 2009), 17.

[13] Organisation for Economic Development and Co-operation, Social Policy Division, "Public spending on education," January 7, 2010, *See* http://www.oecd.org/dataoecd/45/48/37864432.pdf (accessed October 27, 2010). Data originally sourced from the OECD Education database.

[14] Alison Gregor, "Defying the Downturn, Charter School Construction Grows in New York," *New York Times,* August 19, 2009, B6.

[15] National Alliance for Public Charter Schools, "Schools Overview, 2009-10 National," http://www.publiccharters.org/dashboard/schools/page/overview/year/2010 (accessed July 28, 2010).

[16] Terry M. Moe and John E. Chubb, *Liberating Learning: Technology, Politics, and the Future of American Education* (San Francisco: Jossey-Bass, 2009), 96.

[17] I. Elaine Allen and Jeff Seaman, *Online Nation: Five Years of Growth in Online Learning* (Babson Survey Research Group, 2007), 5.

[18] Educomp SmartClass, "About Us," *See* http://www.educomp.planetvidya.com/Admin/aboutus.aspx (accessed July 28, 2010).

[19] Moe and Chubb, 80.

[20] Paul E. Peterson, *Saving Schools: From Horace Mann to Virtual Learning* (Cambridge: Belknap of Harvard UP, 2010), 231.

[21] Ibid., 250.

Acknowledgements

Like all projects at Plunkett Research, Ltd., the research, writing, editing and production of this book was very much a group effort. I am extremely grateful to the entire Plunkett team for their diligence in manning the myriad steps and processes that enabled the birth of this book while keeping our offices and other projects running on time and our clients happy. In particular: Martha Plunkett, my wife and our Executive Editor for the past 10 years, was the primary editor of this book, and she unstintingly discussed, debated, edited, enhanced and assisted my own efforts for months on end without complaint, and sometimes without much sleep. She has a lifetime of writing and editing experience, and her assistance was invaluable. Michael Esterheld, a Senior Editor at Plunkett Research did an exceptional job of reading and re-reading the text, challenging my thoughts and making sure the message was accurately expressed. He also wrote the excellent Reading Group Guide. Andrew Olsen, another Senior Editor in our offices, poured over every word, fact and page, made ample corrections and suggestions, and painstakingly built the endnotes and index.

Extremely talented, intelligent readers outside of our offices generously read through all or part of the manuscript, offered opinions, caught errors and vastly improved the final results. These include good friends Amy Oleson and Sharon Lewis, my sister Gail Christopher, a researcher and entrepreneur in her own right, and an exceptional young financial expert and friend, Marco Rimassa. Many other friends around the world offered their input, viewpoints and support.

Multiple vital tasks were provided by other Plunkett Research staff members, including: production of videos, cover design and graphics by Geoffrey Trudeau; marketing and communications by Alex Avila and Ashley Williams; web site and database oversight by Wenping Guo, PhD, our China-born director of IT who seems quite at home and happy in Houston; and assistance with numerous areas in our offices from Jill Steinberg in particular, along with many other members of or assistants to the Plunkett staff, including Emily Hurley, Suzanne Zarosky, Addie FryeWeaver, Brandon Brison, Christie Manck, Kelly Burke, Jeremy Faulk, Lucinda Gaines, Leandra Hernandez, Mikhail Reyderman, Austin Bunch, Tanmay Waugh and a long list of others—

without you, our work wouldn't have gotten done.

This book relies greatly on facts generated by the superbly talented analysts, researchers and economists at The World Bank, the U.S. Bureau of the Census, the U.S. Federal Reserve Bank, McKinsey & Company, the World Trade Organization, the United Nations, The U.S. Energy Information Administration, and the Organisation for Economic Co-operation and Development (OECD), among many other organizations and leading universities worldwide. I gratefully acknowledge their continuing efforts to keep the world well informed on a wide range of vital social and economic topics.

Next, a very long list of friends and associates have been supportive, sympathetic and helpful in thousands of ways. In particular, I thank David Cooley, Ryan Maurer, Bhavya Sehgal and the rest of the Evalueserve team for insights, friendship and hospitality both in the United States and in India; Patrick Spain for his friendship in America and for introductions to fascinating people in India; Anu Shastri, Ranjit Shastri, P. K. Jain, and Kailash Balani for friendship, introductions and warm hospitality in India; Xin Cao, Chen Liu, David Zhao, Frank Wang and Cherry Huang for introductions and excellent hospitality in China; Scott Davidson for introductions, friendship and hospitality in Singapore and India; Steve Wells for introductions throughout Asia; and Lee Pit Teong and his extensive iGroup team for exceptional hospitality and introductions to people and business practices in Thailand, Singapore, Hong Kong, China, Taiwan, Korea, Japan, Malaysia, India, Indonesia, the Philippines, Vietnam and elsewhere in the Asia-Pacific region.

Finally, thanks to friend and fellow cyclist Marcy Johnson for introductions at M.D. Anderson Cancer Center, and my warmest wishes and eternal gratitude to the staff at the Loma Linda University Medical Center in California: Carl, Sharon, Janice, Lynn, Sitha, Rob, Becky, Nancy, John, Mark and many, many more, as well as to the long list of friends and former patients who introduced me to and helped me through the proton radiation process at LLUMC. I offer my special gratitude to Larry and Karen in this regard. Last names are not given, for protection of their privacy, but that does not diminish my deepest possible thanks.

Index

B

C

J

M

O

P

Q

R

S

U

Z